D0859940

THE
MAD
SCIENCE
BOOK

THE MAD SCIENCE BOOK

100 Amazing Experiments from the History of Science

Reto U. Schneider

Translated by Peter Lewis

Quercus

Contents

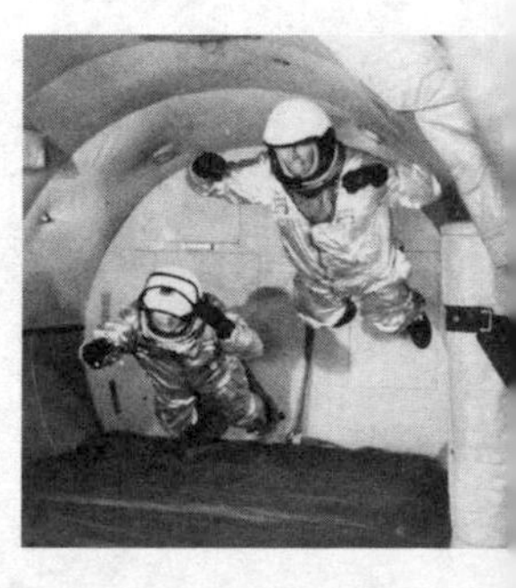

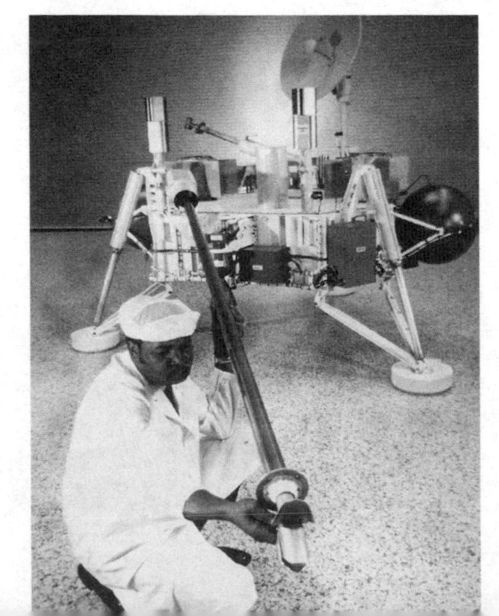

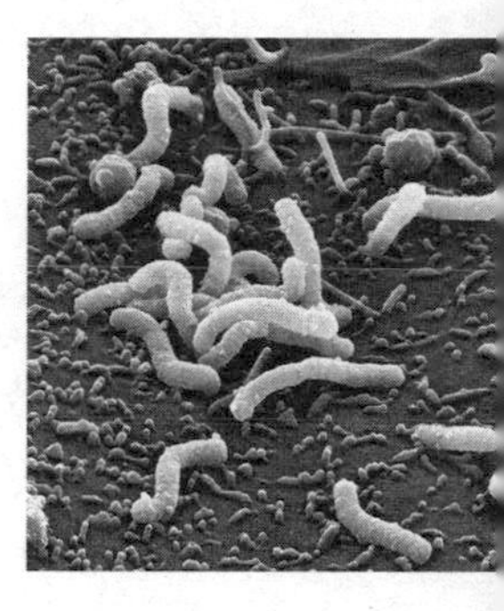

Introduction

This book originated as a by-product. During my time as the head of the science section of a now-defunct Swiss news magazine, I accumulated a stack of research studies about weird experiments. Unfortunately, my editor had no desire to see these appear in print, because they violated all the basic journalistic criteria: they were utterly inconsequential, hopelessly ancient or both.

But since I always considered 'news' an overrated concept in science journalism anyway – after all, Newton is news to most people – I decided to hold on to my pile of clippings. Several years later I was offered a chance to write a science column for *NZZ Folio*, the magazine section of a major Swiss daily newspaper. As my articles didn't have to be related to current events and didn't need to be important in the traditional sense, I at last found myself in a position to write about such subjects as the elixir of life being extracted from guinea-pig testicles or the first meeting between a robot dog and a real dog. The column – several instalments of which are reproduced here – soon had a loyal following. Female readers alerted me to hitchhiking tips; male readers asked for precise details of striptease research in Las Vegas.

The question people asked me most frequently was: 'Where on earth do you find all these strange studies?' But in actual fact a far more interesting question turned out to be how not to find them: don't ask the scientists. Trust me, I tried. The usual answer was something like 'Nothing weird goes on in my field of research'. And if I later confronted them with the strange studies I had unearthed, they gazed at me blankly, unable to see the humour in the psychology of car-horn honking or the economics of restaurant tipping.

The fact that most experiments in this book seem odd by no means implies that they are worthless (though there's no denying that some of them genuinely are). Others appear ridiculous only at first glance, but are in fact truly ingenious. When, in 2005, the German edition of *The Mad Science Book* became a bestseller, some researchers proudly announced on their websites that their experiments were in it.

The Mad Science Book is as much about the informal process of experimentation as it is about formal publication in science journals. I relied on background material, unpublished data, and newspaper articles, and spoke personally with researchers whenever possible. In the process, I stumbled across experiments that destroyed marriages and ended careers, experiments that made headlines and others that have become the stuff of urban myth, although they never actually took place. I have also come to the conclusion that this collection of curiosities may in fact tell us more about the nature of science than any number of reports on cutting-edge research.

Scientific publications often give the impression that experimentation is a linear process: researchers read the relevant literature, construct a hypothesis, and design a test that then runs like clockwork. But, as a scientist once explained to me – and as the reader of this book will soon find – in real life doing an experiment is rather like going to war: 'On first contact with the enemy, all plans go out the window.'

Reto U. Schneider
May 2008

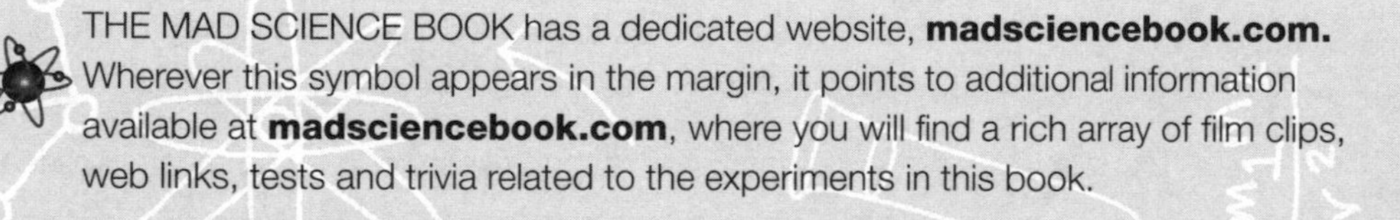

THE MAD SCIENCE BOOK has a dedicated website, **madsciencebook.com.** Wherever this symbol appears in the margin, it points to additional information available at **madsciencebook.com**, where you will find a rich array of film clips, web links, tests and trivia related to the experiments in this book.

1304 Theodoric and the Rainbow

Sometime between 1304 and 1310, the Dominican monk Theodoric von Freiberg filled a spherical glass vessel with water and held it up to the sun. This simple act set in train what would later be regarded as 'the greatest scientific production of the West in the Middle Ages'.

Many scholars before Theodoric had tried to fathom the mystery of the rainbow. Some thought that the arc in the heavens was a reflection of the sun's disk, whereas others took the view that rain clouds acted as a lens. It was obvious that rain somehow reflected sunlight, since the rainbow was only visible when there was a low sun shining from behind the observer. But why did the arc invariably take the form of a segment of a circle with an unchanging diameter? How was the sequence of colours to be interpreted? And what was the origin of the second arc that sometimes appeared above the first and in which the colour sequence was reversed?

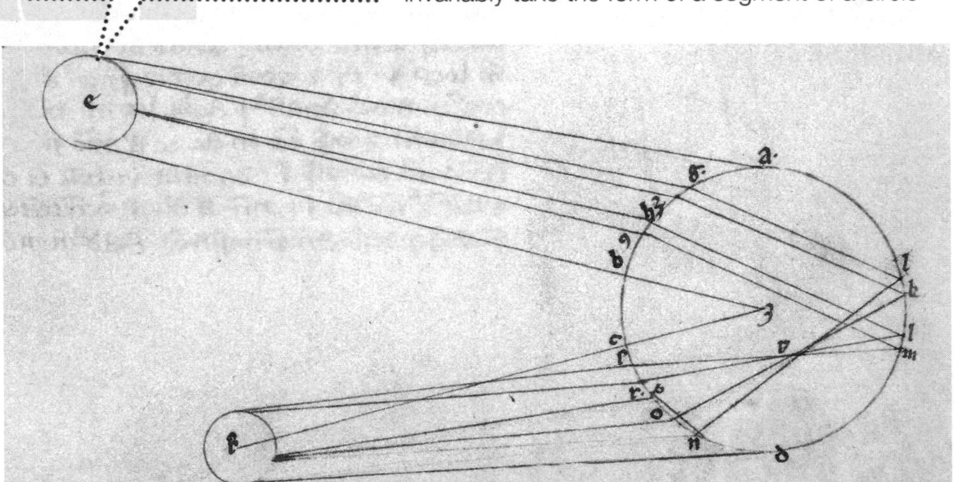

The creation of the primary rainbow, as sketched by Theodoric von Freiberg. The light of the sun (top left) is refracted as it enters the water droplet (right), reflected off its rear surface, refracted again on exiting the droplet and finally reaches the eye (below left) split up into different colours.

Nothing useful was to be gained from simply gazing at the rainbow, but then again, how could this natural phenomenon possibly be brought into the laboratory for study? Although people already knew that sunlight split into different colours when it shone through a flask filled with water, the flask, which was conceived as being a rain cloud in miniature, still didn't produce a rainbow.

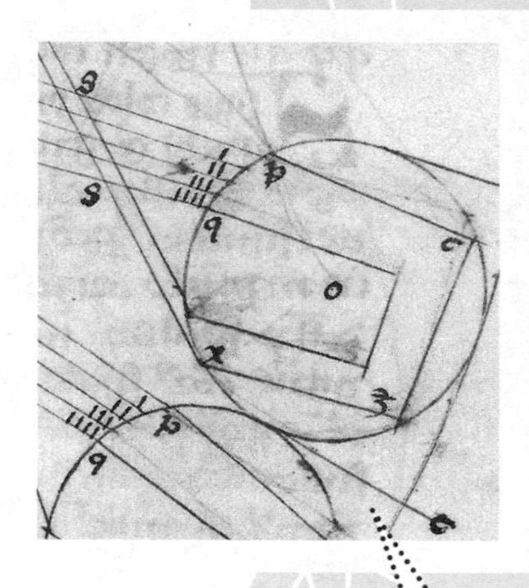

Unlike in the primary rainbow, in the secondary rainbow the ray of sunlight (second line top left) is reflected twice from the rear surface of the water droplet before it exits (five lines).

Clearly, a whole new way of thinking was called for, and Theodoric von Freiberg was the man for the job. His innovation was to view the spherical glass flask not as a miniaturized rain cloud but rather as an enlarged raindrop. Anyone who could get to the bottom of how sunlight behaved when reflected in a single raindrop would then easily be able to deduce from this what occurred when the myriad droplets of water in a shower of rain combined to produce this same effect simultaneously.

And so von Freiberg decided to trace the path of an individual beam of sunlight. First, he arranged it so that the ray of light shone into the upper part of the raindrop. Using the model of his spherical flask, he noticed that the beam first refracted before continuing its passage through the water at a somewhat steeper angle. At the rear of the flask, part of the beam came through the glass, while another part was reflected back, travelling through the water in the opposite direction before emerging near the base of the flask on the side nearest the sun, at which point it was refracted once more.

From other experiments he had conducted, von Freiberg knew that sunlight was split into various colours when it passed through water or glass. Each individual drop therefore always radiated all the colours simultaneously in different directions. Out of these, we only ever perceive at any given moment that colour whose concentrated reflection strikes our eye. Thus, as a raindrop falls, the first part of the spectrum that we see is red, which is reflected at an angle of around 42 degrees, followed by orange, yellow, green, indigo and finally violet, which has a reflective angle of around 41 degrees. A rainbow is therefore made up, so to speak, of a whole series of falling mirrors – droplets of rain, that is – which flash one after the other in all the colours of the rainbow. Because the droplets that have already fallen past the light are constantly replaced by new ones, this gives the impression of a static band of colour.

But what gave rise to the second, somewhat larger rainbow that often appeared above the first? Von Freiberg found the answer by following the path of the ray of light that penetrated to the base of

Source

von Freiberg, D. (around 1310), *w.* Translation: 'Theodoric of Freiberg: On the Rainbow'. In: Grant, E., ed. (1974), *A Source Book in Medieval Science*, pp. 435–441.

madsciencebook.com An interactive animation showing how a rainbow arises.

the flask. The beam was split once more, traversed the water back to the rear of the flask, but was reflected there at such an obtuse angle that after a short passage through the water it hit the rear surface of the flask again. Reflected again, it came out through the top of the flask on the side nearest the sun, where it was reflected back down once more. In other words, a double reflection was involved in producing the secondary rainbow, hence the inversion of the sequence of colours in the primary arc, which was created by just a single reflection. Moreover, the fact that the secondary rainbow is always fainter than the primary can only signify one thing: more light is lost in the course of two reflections than in one.

Even so, Theodoric von Freiberg got one point wrong in his explanation. He thought that the appearance of the separate red, yellow, green and indigo that he saw in the rainbow was dependent upon the depth to which the light ray penetrated and upon the clarity of the water. It was only later that researchers discovered that the colours are actually created by the different wavelengths of refracted light.

Von Freiberg's experiment was one of the first in the history of science. The method he employed – namely that of deducing the properties of individual constituent elements from the properties of the whole – became enshrined as reductionism, the most successful scientific principle of all time. And yet no sooner had he done his experiment than critical voices were heard decrying inquirers into this phenomenon for 'destroying the inherent poetry of rainbows'.

1600 A Balanced Existence

If the *Guinness Book of Records* had existed in his day, then Sanctorius Sanctorius would surely have warranted an entry in it – for no person can ever have spent longer on a pair of scales than the renowned doctor from Padua. His desk, chair and bed were all suspended from ropes that led to concealed counterweights in the ceiling. This was the apparatus that Sanctorius diligently used to record the minutest changes in his weight over a period of 30 years. In addition, he weighed the food that he consumed and the excrement he evacuated.

Source

Sanctorius, S. (1614), *De Statica Medicina.* Translation: Osborn, J. (1728), *Being the Aphorisms of Sanctorius*

He published the conclusions concerning the functions of the human body that he drew from these measurements as a series of pithy maxims in his work *De Statica Medicina* ('On the State of Medicine'), which is nowadays regarded as a classic. The most famous of these concerned the remarkable fact that what a person excretes as urine and stools weighs only a small fraction of what they ate: 'If eight pounds of meat and drink are taken in one day, the quantity that usually goes off by insensible perspiration in that time is five pounds.' Sanctorius had no idea that this 'insensible perspiration' consisted mainly of sweat, yet he was still the first person to measure the quantity, and in so doing became the founder of quantitative experimental medicine. Up until then, doctors had simply taken a descriptive approach to their subject.

Unfortunately, Sanctorius never described his experiments in detail. It is therefore left to the reader's imagination to wonder what the experimental apparatus for maxim number two in the chapter entitled 'On Sexual Intercourse' might have looked like: 'During excessive sexual intercourse, around one-quarter of the normal quantity of perspirations are obstructed.'

In Sanctorius' house, everything hung on scales: the bed, the desk, and – as shown in this copperplate engraving – his chair.

1604 Stones in the Head

Is it possible to refute the mistaken idea that a heavier stone falls at a faster rate than a lighter one without ever holding a stone in your hands? The Italian scholar Galileo Galilei did precisely that in a thought experiment he devised in the early 17th century. At that time, the 2000-year-old opinion of the Greek scholar Aristotle still held sway – namely, that the velocity of free-

madsciencebook.com The Galileo Project contains exhaustive information on Galileo and his age.

falling bodies is proportional to their weight.

In his theoretical experiment, Galileo thought of the heavy and the light stone as being joined together and asked himself how fast the stones would fall in such a case. If Aristotle was indeed right and the heavy stone, on its own, fell faster than the lighter one, then 'the slow stone would decelerate the fast one, whereas the fast one would accelerate the slow one. Their combined velocity should, then, be somewhere between that of the slow stone and that of the fast one.' On the other hand, Galileo argued, the two stones together had to be heavier than the heavy stone alone, and must therefore fall faster than the heavy stone. Aristotle's principle therefore produced a contradiction that could only be resolved by assuming that a body's fall velocity is independent of its weight. The everyday experience that a leaf falls more slowly than a lead shot has nothing to do with the weight of the two objects, but rather with the varying degree of resistance that their shapes and surfaces offer to the air. Lest anyone should still be sceptical, this theory was put to the test on the Moon (see p. 223).

Source

Galileo Galilei (1638), *Discorsi e Dimostrazioni Matematiche Intorno a Due Nuove Scienze appresso gli Elzevirii*. Translation: *Dialogues Concerning the Two New Sciences* (1914). Macmillan.

1620 Wood from Water

'I took a pot, which I filled with 200 pounds of earth that I had dried in an oven and then moistened with rain water, and planted in it a willow sapling weighing five pounds.' This brief description of method is the starting point for one of Johan Baptista Van Helmont's experiments.

The Flemish scholar could not have known that it would become his most famous experiment; after all, he had already conducted many more spectacular tests. For example, he had once (reputedly) changed a pound of mercury into eight ounces of gold, and on another occasion he was convinced that he had found the recipe for creating life: 'If you stuff a dirty shirt into a hole in a barrel that has been filled with grains of wheat, after about 21 days, there will be a noticeable change in the smell and the products of decomposition will penetrate the husks of the wheat and transform the grains into mice.' When Van Helmont carried out this latter experiment, he was less surprised by the actual appearance of the animals than he was by the fact that mice of both sexes were produced.

Van Helmont was history's last alchemist and first chemist, and his view of the world was a blend of magic and science. He was born in Brussels in 1579 to a prosperous family and, after dabbling in almost every subject area, graduated in medicine from the University of Louvain in 1599. Shortly afterwards, he withdrew from public life and became a freelance scholar.

In his laboratory, he investigated the properties of gases, observed the fermentation of various materials and devised new medicaments. It is not known when precisely he took up his shovel and hoe for the willow-tree experiment. The first written record of it dates from 1648, four years after Van Helmont's death, when his son published his father's collected works in the volume *Ortus medicinae* ('The Origins of Medicine').

In this book, Van Helmont also expounded his philosophy of nature, which the experiment with the willow tree in the pot was designed to corroborate. In contrast to the Greek philosopher Aristotle, who maintained that all matter is composed of the four elements earth, water, fire and air, Van Helmont regarded only two of these as true elements, namely air and water. Fire could produce nothing on its own, while earth was inferior to water in its purity and simplicity. Moreover, in the history of creation, water even made an appearance before the first day. Van Helmont was convinced that all matter – be it rocks, earth, animals or plants – was ultimately made up of water. The experiment was designed to prove this hypothesis where plants were concerned.

Five years after he had planted the willow, he uprooted it from the earth and weighed both: over this period, just two ounces (56 g) of the earth had gone missing, while the tree had, by contrast, grown to a weight of 169 lb 3 oz (77 kg), thereby gaining more than 30 times its original weight.

From these findings, Van Helmont drew the only reasonable conclusion given the state of scientific knowledge at that time: 'One hundred and sixty-four pounds of wood, bark and roots,' he observed, 'have been produced by water alone.' For, aside from watering it regularly, he had left the tree to its own devices.

Van Helmont knew that the result was hardly surprising. Long before he came on the scene, scholars had postulated the same test as a thought experiment, and had obtained the same outcome. However, he was the first to conduct the experiment for real, using earth, a tree and scales, and in so doing had paved the way

madsciencebook.com Animation showing the process of photosynthesis.

Source

Van Helmont, J. B. (1648), *Ortus medicinae*. Elzevir. Translation in: *A Source Book in Chemistry, 1400–1900* (1952). McGraw-Hill.

for the scientific experiment to become an indispensable tool for advancing knowledge.

Stimulated by Van Helmont's ideas, other researchers were soon busy carrying out experiments with pot plants in their own laboratories. These investigations revealed that the pioneering Belgian had not been quite right in his conclusions: plants didn't just require water to grow, but also air, light and trace elements of minerals from the earth.

Van Helmont's experiment was an important first step on the road to explaining the mysterious process that would later become known as 'photosynthesis': that is, the conversion, by means of light, of the energy-poor compound of water and carbon dioxide into energy-rich compounds that provide animals with sustenance. Without noticing at the time, early scholars hit on the most important difference between plants and animals: only plants are capable of storing energy from the sun in this way, in the form of chemical compounds. All animals – including man – are directly or indirectly dependent upon this process.

Van Helmont's experiment was revived in the 20th century. Students were presented with it to test their mental agility and used it as a model for practising the design of simple, elegant experiments. Practice exercises relating to it can even be found on the Internet. So as not to prolong the student's education unnecessarily, however, they are encouraged to grow radishes rather than willow saplings.

1729 The Mimosa Clock

When he put one of his pot plants away in a cupboard, the French astronomer Jean-Jacques d'Ortous de Mairan had no inkling that he was founding a new branch of science. Indeed, so inconsequential did his mimosa experiment seem to him that he didn't even bother publishing the findings.

Mimosas close their leaves at night and open them during the day. De Mairan asked himself what would happen if the mimosa no longer knew whether it was night or day. And so, at the end of the summer of 1729, he placed a Mimosa pudica plant in a pitch-black box and discovered that the leaves opened and closed at the

Source

De Mairan, J.J.D. (1729), 'Observation Botanique'. *Histoire de l'Académie Royale des Sciences*, p. 35.

right times even in the absence of sunlight. As a friend of de Mairan and Académie Française member put it in a letter to the highest scientific committee in France, the Académie Royale des Sciences: 'Thus, the mimosa feels the presence of the sun without seeing it.'

This conclusion was wrong. Much later, researchers determined that the mimosa does not feel the presence of the sun but rather has an inbuilt clock. Nevertheless, de Mairan is regarded today as the founder of chronobiology, the field of science that studies the internal rhythms of living organisms. Two hundred years later, a scientist repeated de Mairan's experiment – but this time, rather than using plants, he and his assistant spent a month in a darkened cave (see p. 107)

The astronomer Jean-Jacques d'Ortous de Mairan put a mimosa plant in a dark cupboard, so founding a new branch of science.

1758 The Philosopher's Socks

'I had for some time observed, that upon pulling off my stockings in an evening they frequently made a crackling or snapping noise.' These are the opening words of the English scholar Robert Symmer's article in *Philosophical Transactions,* the leading scientific journal of the period. Since Symmer's friends had had similar experiences with their own hosiery and yet he knew of nobody who had examined this phenomenon 'in a philosophical way', he had, he claimed, found himself obliged to undertake 'the strictest inquiry possible' into the matter.

This claim was by no means an exaggeration. Symmer presented his findings at no fewer than three meetings of the Royal Society. The detailed reports he prepared on his exciting encounters with his socks ultimately filled over 30 pages – later prompting the French to give him the nickname 'the barefoot philosopher' (*philosophe déchaussé*).

There was no shortage of opportunities for observation, for, as Symmer explained, 'the simplicity of the apparatus' – his socks, that is – and the 'great facility in making the proper

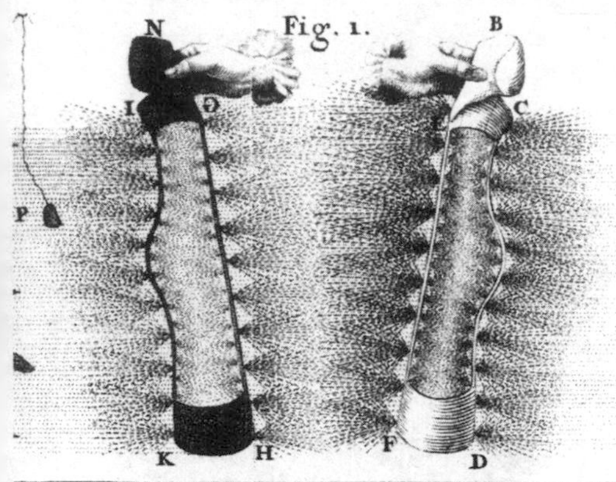

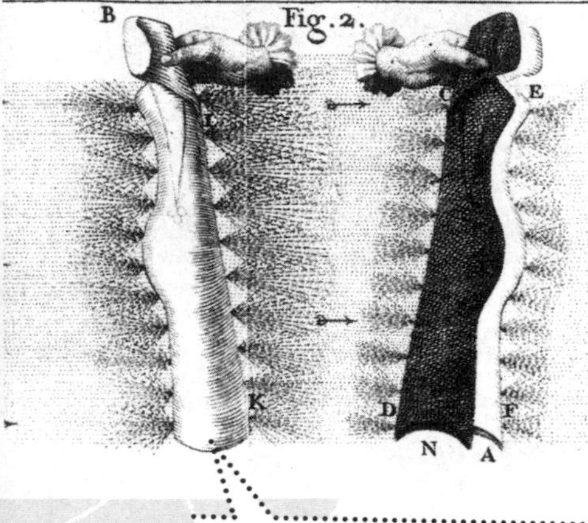

These experiments with his socks earned Robert Symmer the nickname of the 'barefoot philosopher'.

experiments' – by which he meant the act of putting his socks on and taking them off again – 'put it in my power to begin and carry my enquiry at pleasure'. After a few tests with socks made variously of cotton, wool and silk, the first thing that Symmer discovered was that a combination of wool and silk suited his experiments best. It made no odds whether he wore the woollen socks over the silk ones or vice versa: the most important thing was to take them off together and only then to separate them. As he did so, they became electrically charged. Symmer could tell this from the fact that the socks filled out, as if they were being held in a breeze, and because they were attracted to one another when they were held close together.

In a second series of experiments, Symmer used only one black and one white silk sock, because these produced the strongest effect. He also changed his modus operandi. 'Having found it troublesome to electrify the stockings, by putting them as often on my legs as was requisite in making experiments, I have quitted that method entirely, and satisfy myself with the degree of electricity which is excited in the stockings by drawing them upon the hand.' Symmer further maintained that this had the advantage that the socks remained usable longer for the tests, for 'like other electrical apparatus, they must be kept clean'.

Symmer was well aware that some people were laughing up their sleeves at his experiments, and could even sympathize with their attitude – to some extent. As he wrote to a friend: 'You may likewise be disgusted with the frequent mention of pulling on, and putting off, of stockings: a circumstance, I confess, so little philosophical, and so apt to excite ludicrous ideas, that I was not surprised to find it the occasion of many a joke, among a sarcastical set of minute Philosophers, who do not love to have anything new forced on them.'

Source

Symmer, R. (1759), 'New Experiments and Observations Concerning Electricity'. *Philosophical Transactions of the Royal Society* 51, p. 340.

1772 Electrocuting Eunuchs

Not long after the invention of the Leyden jar, a device that allows electrical charges to be stored, a singular rumour began to do the rounds in Paris. At that time there was something of a craze for using this apparatus to electrocute long chains of people who had joined hands. Ladies and gentlemen from the higher echelons of society would shriek with pleasure as an electric shock made 20 people jump simultaneously. Even the king submitted himself to a demonstration of the remarkable electrical phenomenon. This parlour game was repeated using 180 soldiers and later 200 Carthusian monks (or possibly even 700, as some sources claim). Yet during some demonstrations, something unexpected occurred: the electric shock fizzled out halfway down the line.

For example, when the scholar Joseph Aignan Sigau de la Fond attempted to electrocute 60 people in the courtyard of a school in Paris, try as he might, he could never get the shock to travel beyond the sixth person. Because the young man standing in this position was suspected of 'not being equipped with everything that makes a man a man', the rumour arose 'that it was impossible to electrocute those whom nature had cursed in this way'.

Although Sigau de la Fond scoffed at the idea, when he was summoned to the royal court to give a demonstration that would test the hypothesis, he jumped at the chance. The guinea pigs in this case were three of the king's musicians 'whose condition was in no doubt' (i.e. they were castrati). And, indeed, Sigau de la Fond was proved right: at no point in the chain did the royal eunuchs break the electrical circuit. Quite the opposite, in fact – they appeared to react particularly sensitively to the electric shocks.

'And so,' the German physicist and philosopher Georg Christoph Lichtenberg later wrote, 'the electricity-generating machine lost out on the honour of taking pride of place in the assembly rooms of the consistories and marriage courts'.

The real reason why electricity did not travel equally easily through all human chains was not to do with certain men's impotence (or, as was also suspected, women's frigidity), but with the conductivity of the ground they were standing on. For

Source

Sigaud de la Fond, J.A. (1785), *Précis historique et expérimental des phénomènes électriques: depuis l'origine de cette découverte jusqu'à ce jour*, p. 231.

instance, if it was damp, then a large amount of the electricity simply earthed down the participants' legs and failed to reach the limbs of people further down the chain.

1774 A Sauna in the Cause of Science

On 23 January 1774 the physician Charles Blagden was invited by his colleague George Fordyce to take part in a series of experiments. What the two men then proceeded to do in the service of science was scarcely any different from what millions of people now do every week for their own well-being and health: they went to a sauna. Except that this sauna visit was the best-documented in the annals of human history.

The physician George Fordyce had a kind of sauna built for his experiments with heat.

Over 24 pages in the Transactions of the Royal Society, Blagden publicly recounted his and the other participants' experiences in the steamy heat. In addition to Blagden and Fordyce, the experiment also involved the Hon. Captain Phipps, Lord Seaforth, Sir George Home, Mr Dundas, Mr Banks, Dr Solander (who sweated more profusely than anyone) and Dr North.

In all likelihood, Fordyce had no idea that the room he had specially constructed for this experiment was, to all intents and purposes, a sauna. It consisted of three rooms, the hottest of which had a domed roof and was heated in two ways: firstly by hot-air ducts in the floor, and secondly by Fordyce's servants dousing the external walls with buckets of hot water.

The purpose of this edifice was to help the researchers discover how high a temperature the human body can endure. They began with a modest 45°C but soon cranked it up to 100° (210°F), and then to 127° (260°F). At first, they would sweat for eight minutes fully clothed – in their outdoor clothes, including gloves and stockings – but later sat there naked, at which point they also introduced a frying pan containing a steak.

After three-quarters of an hour, the meat was 'not only dressed, but almost dry,' as Blagden later wrote. With the next steak they tried, just 33 minutes passed before Blagden lifted it out of the pan and pronounced it 'rather overdone', while the third – which Blagden helped along by stirring up the air in the heated room with a pair of bellows – was cooked through after only 13 minutes. In and of themselves, these findings weren't especially remarkable – after all, the steaks were being heated to a temperature of over 100 degrees. But what really astonished Blagden was that the sweltering heat had no adverse effect on him whatsoever.

Dead meat was done in no time, but under the same conditions a living and breathing person left the room completely unharmed. Blagden's conclusion was that living organisms had the unique ability to destroy heat. In claiming this, he wasn't referring, say, to the cooling of the human body through sweating, but rather to a 'provision of nature which seems more immediately connected with the powers of life'.

Blagden was wrong to draw this conclusion: there is no such vital force capable of destroying heat. The cooling of the body is achieved solely by the evaporation of moisture such as sweat or spittle in conjunction with the expansion of blood vessels.

Source

Blagden, C. (1775), 'Experiments and Observations in a Heated Room'. *Philosophical Transactions of the Royal Society* 65, pp. 111–123, 484–495.

1783 Where Sheep May Safely Fly

It will forever remain a mystery why this experiment was not conducted much earlier. Its theoretical basis, in terms of physics, had been known about for 2000 years. The wherewithal to construct the necessary equipment had also been available for some time. And yet it was only on 19 September 1783 that the first passengers made an ascent in a hot-air balloon: they comprised a duck (a creature of the water), a sheep (a land-based creature), and a cockerel (a creature of the air).

It had been just a year since the brothers Joseph Michel and Jacques Étienne Montgolfier had undertaken their first experiments with balloons at Annonay, south of Lyons. Various more or less plausible accounts have been offered as to how Joseph Montgolfier first hit on the idea for his experiments: according to one, what triggered his interest was his wife's slip hanging up over the chimneypiece to dry and billowing out in the warm air; another version claims that it was a paper bag carelessly tossed into the fire, which was then carried aloft; yet others maintain that it was simply the sight of rising smoke, or of clouds. But whatever the reason, Joseph Montgolfier found himself directing his efforts towards 'enclosing a cloud in a bag and letting the latter be lifted up by the buoyancy of the former', by channelling the smoke from a fire into a large paper bag. His brother Étienne was enthralled by the idea and helped him assemble his next, larger balloon.

News soon reached King Louis XVI about the curious flying machines being made by the Montgolfier brothers – who came from a family of paper manufacturers – and he invited them to give a demonstration at Versailles. Because the balloon they prepared specially for this occasion was destroyed in a thunderstorm, a new one had to be built in the space of just a few days. Using long bolts of cotton, Étienne Montgolfier and his assistants carefully cut out and stitched together an envelope that was shaped like a ball with a tube attached, and then lined the whole thing with paper. The hot-air balloon was brought to Versailles at 8 o'clock sharp on the morning of the demonstration and carried to the dais that had been specially erected for filling and launching the craft.

Montgolfier was convinced that he had discovered the ideal gas for making a balloon rise: the evil-smelling smoke from a fire. He could not have known that smoke and smell had nothing to do with lift. In actual fact, it is heat that causes the air to expand and thereby become lighter than the same volume of cold air.

At 12 o'clock Montgolfier ordered the fire under the dais to be lit. Given the fact that 80 pounds of straw, 5 pounds of wool and, in particular, old shoes and rotting meat were being burned, the royal hosts chose to observe proceedings from a safe distance.

'The craft was filled in four minutes,' Étienne later wrote to his brother, who had stayed behind in Annonay, 'everyone let go of the guy ropes simultaneously, and the balloon rose majestically into the air. Just after it had started to climb, though, a gust of wind made it pitch sideways. At that moment, I feared disaster might strike.' Yet the 60-ft (18-m) high

Source

Cavallo, T. (1785), *The History and Practice of Aerostation*, p. 68.

balloon righted itself and carried the sheep, the rooster and the duck in their wicker basket up to a height of 1444 ft (440 m).

19 September 1783: the world's first air passengers – a sheep, a rooster and a duck – make their ascent from Versailles in a hot-air balloon.

At first, the thousands of onlookers who had gathered on the square at Versailles simply gazed in astonishment at the flying machine, before breaking out into spontaneous cheering. Eight minutes after lift-off, the balloon came to earth gently some 2 miles (3 km) away. As the craft brushed against a branch, it tore open the wicker basket, releasing the animals from their confinement. The sheep was found grazing peacefully in a nearby meadow, and the duck was also unharmed. Only the cockerel gave rise to some speculation: its right wing was injured, leading concerned onlookers to ask whether man himself would ever dare venture into the skies. However, witnesses soon reported that the cockerel's injury was 'the result of a kick it had received from the sheep at least half an hour before'.

One month later, on 15 October 1783, the first human being made an ascent in a balloon.

1802 The Winking Corpse

On a cold January day in 1802, Giovanni Aldini was waiting in his rooms just off the Piazza della Giustizia in Bologna for the subjects of his experiment to arrive. On a large wooden table scalpels, saws and

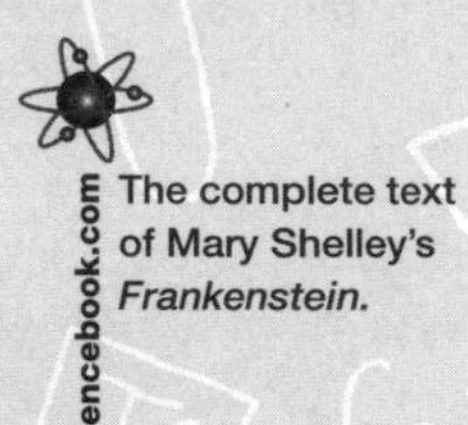

madsciencebook.com The complete text of Mary Shelley's *Frankenstein.*

cables all lay ready, and next to them a strange knee-high column – the voltaic pile. Made up of 100 zinc discs, and the same number each of silver discs and discs of leather soaked in brine, all piled up in alternating layers, this was the first device that was able to supply a constant electric current, thereby facilitating the study of electricity. Yet Aldini knew nothing of this. Indeed, working as he was a full century before the electron was even given its name, how could he have been expected to realize that what he actually had before him was a battery, which set electrons in motion and in so doing generated an electric current?

Being more a showman than a scientist, Aldini was primarily interested in practicalities: with the 100 volts generated by the pile, he had already made a severed bull's head blink. Now he wanted to try out the effect of electricity on a 'more worthy subject'. His experiments required a 'human cadaver with as much life force remaining in it as possible'. Such an item could only be had from one place: the scaffold.

Giovanni Aldini was the nephew of Luigi Galvani, who in his numerous experiments with frogs' legs had observed that their muscles twitched when they were touched by two different types of metal that were in contact with one another. Galvani thought that 'animal electricity' was responsible for this effect, lying dormant within the frogs' legs and being released by contact with the metals. He could not have known that the exact opposite was the case: his apparatus was in fact a primitive battery that electrified the frogs' legs. Because the frogs' legs moved as though they were still alive, Galvani assumed that 'animal electricity' was bound up with the force inhabiting living organisms, and that it was different from the electricity that was found in inanimate matter. However, Alessandro Volta, who invented the voltaic pile in 1800, believed that there was only one kind of electricity, which was responsible both for producing lightning during thunderstorms and for making the frogs' legs twitch.

Galvani's experiments spelt bad news for frogs. Throughout Europe, wherever people could lay their hands on frogs' legs and two different types of metal, scholars and amateur researchers alike began to play around with them. One such researcher was the Scottish physician James Lind, whom the young Percy Bysshe Shelley often visited in his laboratory. Shelley was later to become the husband of Mary Shelley, author of the novel *Frankenstein* – in which

Source

Aldini, J. (1804), *Essai théorique et expérimental sur le galvanisme. L'Imprimerie de Fournier fils*, p. 121.

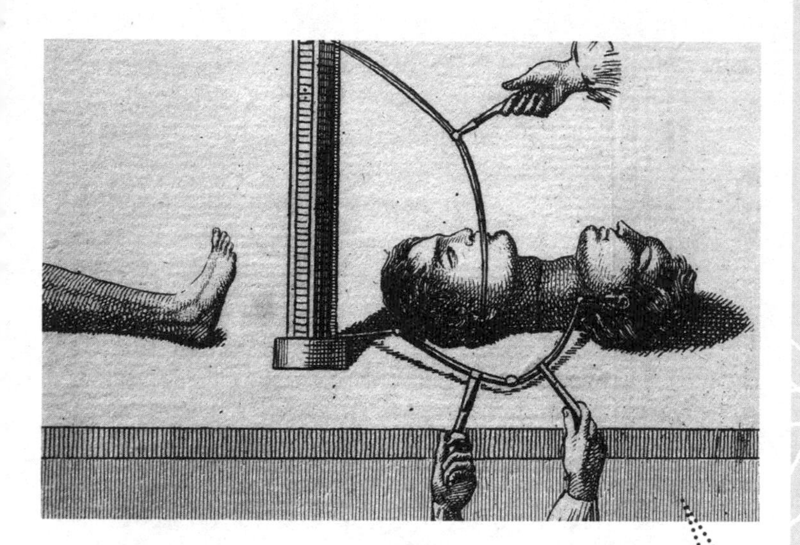

It was a frequent occurrence for spectators at experiments involving electricity and beheaded corpses to faint. Giovanni Aldini's demonstration in Bologna, in which he used a voltaic pile to make these two heads pull grimaces, was no exception.

a creature assembled from body parts taken from corpses is brought to life by electricity. The experiments with frogs' legs that abounded at this period may well, then, be the inspiration for one of the best-loved books in world literature.

The first corpse was delivered to Aldini three-quarters of an hour after the execution. He laid its severed head on the table, inserted a cable into each ear and noticed 'violent twitching in all the facial muscles, which tensed up in such an irregular fashion that they produced the most grotesque grimaces'. Aldini then experimented further by taking one of the wires and placing it first in the corpse's mouth and then in its nose. He shaved its head, sawed its skull open, and poked around in the brain. And just as he was in danger of running out of ideas as to where else he might stick an electrode, he was presented with a second head.

Aldini placed the two heads together at the point where they had been severed and charged them with electricity. 'The grimaces that the two faces pulled at one another were both remarkable and alarming,' he later wrote in his book *Essai théoretique et experimental sur le galvanisme*, adding that the sight caused a number of onlookers to faint.

This spectacle didn't yield much in the way of new knowledge. After describing his fortieth and final such demonstration, Aldini concluded that further experiments would be necessary to shed light on the true nature of galvanism.

He was fully aware that his electric circus featuring execution victims wasn't to everyone's taste. Between descriptions of his macabre experiments, he was at pains to stress repeatedly the honourable motives that induced him to overcome his revulsion at the idea of conducting experiments with beheaded corpses: namely, his love of truth, of humanity, and of science.

He regarded experiments with hanged men as immoral, because their bodies were still in one piece. Anyone carrying out such experiments automatically became a 'barbaric experimenter'. Admittedly, he only reached this conclusion after conducting a series of 14 experiments of his own on the body of the murderer Thomas Foster, who was hanged in London on 17 January 1803.

1802 A Disgusting Doctorate

Medical students who complain about the trouble they're having writing their doctoral thesis should cast an eye on the dissertation submitted by Stubbins Ffirth 200 years ago at the University of Pennsylvania.

Ffirth was just 18 years old when he took it upon himself to prove that yellow fever cannot be transmitted from one person to another. Although chiefly prevalent in tropical regions, this disease also occurred in the southern United States. Its first signs were flu-like symptoms. There then followed three to four days in which the patient experienced a raised temperature, violent shivers, headaches and constant vomiting. The vomit was black, while the skin took on a yellowish hue. In many cases, the illness proved fatal within seven to ten days. Because yellow fever often spread like an epidemic, many people believed that they could catch it from clothes, bedding or other objects that those who had contracted the disease had come into contact with. Indeed, at the outset, Ffirth was also certain this was the case. However, he changed his mind when he realized there was nothing to indicate that nurses, doctors, victims' relatives or gravediggers had a higher incidence of yellow fever than the rest of the population.

Ffirth aimed to demonstrate by experiment that it was entirely safe for people to have contact with yellow-fever patients. He began by feeding a small dog pieces of bread saturated with the vomit of a yellow-fever sufferer. After three days, the dog acquired such a taste for these morsels that it would wolf down the person's sputum even without the bread – and suffered no ill effects. Ffirth then proceeded to feed a cat the same diet, and it too stayed perfectly healthy. And so Ffirth turned his attention to the dog once more; making an incision in the skin on its back, he filled the wound with vomit and sewed it up. Again, the dog remained healthy. It was only when Ffirth injected the sputum directly into the dog's jugular vein that it died. But Firth was convinced that its death had nothing to do with yellow fever, since another experiment proved that a dog died even when its jugular was injected with water.

Finally, on 4 October 1802, Ffirth started to experiment on a new guinea pig: himself. He cut open his forearm and poured vomit from a yellow-fever patient into the wound. Nothing happened. To be on the safe side, he repeated this experiment at around 20 more places on his body. He then dripped vomit into his eyes, put vomit on a fire and breathed in the fumes, swallowed pills made of dried and pressed vomit, swallowed it in diluted form 'and increased the quantity I took from half an ounce (14 g) to two ounces (56 g), drinking it at length without dilution', as he subsequently reported in his dissertation.

Once he was convinced that the disease could not be passed on through vomit, he turned his attention to patients' blood, spittle, sweat and urine. He swallowed their blood 'in considerable quantities' and lacerated himself in a number of places, into which he poured various bodily fluids. Here he was extremely fortunate, as the virus could indeed have been transmitted through the bloodstream. It may either have been the case that Ffirth was already immune by this stage, or perhaps the blood was no longer virulent when he used it. In any event, he did not fall ill, and by now was sure that you couldn't contract yellow fever just by being in the presence of someone infected with it.

Neverthless, Ffirth's heroic experiments made little impact on medicine. The main thing they demonstrated was how yellow fever didn't spread, but the more important question they failed to address was how the disease actually did spread.

Yet Ffirth already recognized the key indicators of the disease. As he wrote in 1804, yellow fever differed from a contagious illness 'in occurring only during the

Source

Ffirth, S. (1804), *On Malignant Fever: With Attempt to Prove its Non-Contagious Nature, from Reason, Observation, and Experiment.* B. Graves.

existence of hot or warm weather, by being checked by cold, and by its never becoming epidemic when the thermometer is below 32°'. A century later, researchers were to find that the virus was transmitted by mosquitoes.

1825 The Man with the Hole in his Stomach

It was on 6 June 1822, shortly after midday, that William Beaumont knelt down next to the bleeding soldier. A musket had accidentally been discharged in the storeroom at Fort Mackinac on the Canadian border between Lake Michigan and Lake Huron, hitting the soldier Alexis St. Martin in the stomach. Beaumont removed splinters of bone and scraps of clothing from the wound, snipped off a piece of broken rib that was sticking into the man's lung, and applied a poultice made from a mixture of flour, hot water, charcoal and yeast. As an army surgeon, Beaumont had considerable experience in treating gunshot wounds, and this was patently a hopeless case. But on this occasion Beaumont's prognosis turned out to be wrong: although the 28-year-old soldier contracted a violent fever and pneumonia and was bled by

The army doctor William Beaumont siphoning gastric juices from Alexis St. Martin through the hole in his stomach. This picture was painted more than 100 years after the actual experiment, as part of a series on the great pioneers of American medicine.

Beaumont, his condition improved. The only hitch was that the wound refused to heal over. Whatever St. Martin ate re-emerged from his body through the hole under his left breast. At first, it was necessary to bind his chest in a tight dressing to prevent this from happening while the patient was eating and digesting, but later a bulge of skin created a fistula. Although just by pressing lightly, you could now poke your finger directly into his stomach, bandages were no longer necessary.

The most famous chest image in medicine: a gunshot wound left a hole in the side of the soldier Alexis St. Martin, which led directly into his stomach.

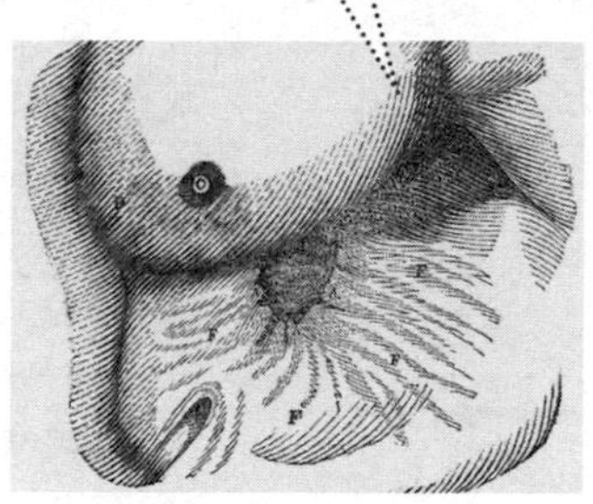

Beaumont later claimed that his motives in caring for St. Martin were entirely unselfish. Even so, he had probably spotted straight away what a golden opportunity the hole in his patient's side presented. In any event, he successfully prevented St. Martin from being removed from his care and posted to Montreal after his long convalescence.

On 1 August 1825, at around 12 o'clock midday, Beaumont stuffed 'a piece of high seasoned alamode beef, a piece of raw salted beef, a piece of raw salted fat pork, a piece of raw lean fresh beef, a piece of boiled corned beef, a piece of stale bread and a bunch of raw cabbage' into the hole in St. Martin's side, all secured on a length of silk thread. At intervals of one, two and three hours, he pulled the bundle out to observe it. This was the first of many experiments. During the second such experiment, Beaumont siphoned off gastric juices into a container, in which he had placed a slice of corned beef. As he watched, the meat was digested.

This answered an age-old question, namely, was digestion a purely chemical process or was some vital force also involved, which made the difference in the human body between digestion and decomposition? Clearly, no such 'vital force' was necessary; the chemical action of the gastric juices in a dish was enough to set the process in motion. Beaumont also compared the effect of saliva and gastric juices and in so doing refuted the assumption that gastric juices were simply saliva that had collected in the stomach. He also ascertained that gastric juices dissolved food faster than diluted acid. Later, another researcher discovered the enzyme pepsin in the stomach, which breaks down proteins.

In September 1825, shortly after Beaumont began his experiments, St. Martin moved to Canada – 'without obtaining my consent,' as Beaumont later emphasized – where he married and had two children. Two years later, Beaumont tracked him down and persuaded him, in return for payment, to resume his life as his patient.

The physician now made a series of observations of the movements that St. Martin's stomach made, studied his stomach lining, and fed him lavish meals, which he then removed from his stomach 20 minutes later. He determined the time it took to digest various different foodstuffs and investigated the effect of the weather on digestion time. His guinea pig had signed a contract that committed him to 'serve, abide, and continue with Beaumont wheresoever he should go' and to 'comply with all reasonable and proper orders or experiments'. In return, he was to receive £150 per annum, plus food and lodging.

The experiments were extremely strenuous. In certain months, St. Martin had to endure experiments on an almost daily basis; in 1832 one of these even took place on Christmas Day. In 1834 he went to visit his family in the south of Canada and never returned. Beaumont died 19 years later aged 68. Until shortly before his death, he attempted to get St. Martin back. One year before his death he wrote him a letter that ended as follows: 'I can say no more, Alexis – you know what I have done for you many years since – what I have been trying, and still anxious and wishing to do withal for you – what efforts, anxieties, anticipations and disappointments I have suffered from your non-fulfilment of my expectations. Don't disappoint me more nor forfeit the bounties and blessings reserved for you.' The case had by then become widely known in the medical world. Colleagues from London and Paris were eager to see the man with the hole in his stomach. Beaumont also lived in constant fear of other physicians making direct contact with his protégé.

As far as Beaumont was concerned, Alexis St. Martin was an experimental subject whom he had the right to dispense with as he saw fit. In common with most contemporary doctors, he had no ethical qualms – either concerning the possible ramifications of his experiments or in view of the fact that he forced St. Martin to abandon his family for years on end. In the foreword to his book *Experiments and Observations on the Gastric Juice and the Physiology of Digestion*, which appeared in 1833 and is now regarded as a classic, he expressed his gratitude to a whole string of physicians for their assistance, whereas the subject of his studies hardly gets a mention.

Alexis St. Martin died on 24 June 1880, 27 years after Beaumont, at the age of 86. A number of doctors

Source

Beaumont, W. (1833), *Experiments and Observations on the Gastric Juice and the Physiology of Digestion*. F.P. Allen.

were keen to perform an autopsy on his body and donate his stomach to a museum. However, his family preferred to lay out his corpse at home until putrefaction set in. Then they buried him. The grave was dug to a depth of 8 ft (2.4 m) to prevent anyone from exhuming his body.

1837 … and Introducing Darwin on Bassoon

There are certain experiments that you just can't help picturing from the viewpoint of the animals being experimented upon. So there is this earthworm meandering its way around a flowerpot filled with earth, and what does it see when it peeps over the edge? None other than one of the foremost natural scientists of all time, Charles Darwin, holding a bassoon right up to the pot and puffing out his cheeks to produce the lowest note he can reach. But you'd be wrong to suppose that this would have taken the worm by surprise; in fact the great scholar had already treated it to recitals on the flute and the piano.

Darwin was not only the originator of the theory of evolution, he also spent more than 40 years doing intensive research on the life of the earthworm. Amongst other things, he wanted to answer the question of whether earthworms could hear. But when they failed to react to any of the instruments and also showed no response whatever when Darwin yelled at them, he concluded in his 1881 book *The Formation of Vegetable Mould, through the Action of Worms* that 'Worms do not possess any sense of hearing.'

Source

Darwin, C. (1881), *The Formation* of *Vegetable Mould, through the Action of Worms.* John Murray, p. 12.

1845 Trumpeters on the Train

It could have been a concert organized specially for Dadaists: the locomotive that spent the day of 3 June 1845 shuttling between the towns of Utrecht and Maasen in Holland was pulling just a single flat-bed truck behind it, on which stood three men. One occupied himself with jotting down figures on a printed form, while another played a G on a trumpet whenever the third man signalled that he should do so.

Standing next to the fireman on the locomotive's footplate, Christoph Buys Ballot cast nervous glances up at the sky and prayed inwardly that the weather wouldn't turn. This 28-year-old physicist had been forced to abandon his first attempt to conduct this experiment in February. On that occasion, the musicians were confronted by a blizzard and the cold had put their instruments out of tune. By contrast, this Tuesday was a mild summer's day, and Buys Ballot stood a good chance of seeing through his experiment. With the help of six trumpet players, two watches and a locomotive, it was designed to test the soundness of a theory that an obscure Austrian professor had devised in 1842 on the colour of stars.

Three years had passed since Buys Ballot had obtained a copy of 'A Monograph by Mr Doppler'. In this work, which was entitled 'On the Coloured Light of the Binary Refracted Stars and Other Celestial Bodies', Christian Doppler postulated that anyone approaching or moving away from a light source at great speed would perceive different colours in it than if he had remained static. This phenomenon could not be observed in everyday life, since it only occurred at high velocities. Nevertheless, Doppler was convinced that anyone seeking to confirm his theory need only look at the stars.

Astronomers had divided the stars that appeared in the night sky into two categories: white stars and coloured stars. White stars were individual stars that appeared not to move, while coloured stars often formed one element of binary stars – that is, two stars orbiting around one another. Doppler believed that the colour of binary stars had to do with the fact that they were alternately moving away from and closer to Earth. The theory that he based on this idea has gone down in the history of physics as the Doppler effect.

Following various disputes concerning the nature of light, physicists in Doppler's period were broadly in agreement that light radiates like a wave and that the different colours it assumes arise from the different frequencies at which light waves oscillate; thus, violet light oscillates fastest, and red light slowest, while in between, as in a rainbow, came blue, green, yellow and orange.

The factor that determines whether a person perceives red or blue is in how rapid a succession the waves of light strike the observer's eye. Doppler was amazed to discover that nobody hitherto had noticed that motion of the light source or of the observer also played a role in this. A person approaching a light source is moving against the direction of the wave and therefore encounters the waves in quicker succession than if he were standing still. Conversely, anyone moving away from a light source is distancing himself from the wave pulses, which now take longer to catch up with him and thus reach him in slower succession. The same principle also holds good for the opposite case, in which the observer remains static and the light source moves.

In his experiments to test the Doppler effect, the physicist Christoph Buys Ballot had to contend with ill-disciplined orchestral musicians.

Doppler provided a graphic illustration of the principle with the example of a ship steering into the waves and 'encountering over the same period a greater number of waves and a more violent pounding by them than a ship that either remained motionless or one that was being carried along by the waves and travelling in the same direction as them'.

In his paper, Doppler also calculated the speed at which this effect would become visible to the naked eye – '33 miles a second'. This figure effectively discouraged even the most optimistic researcher from trying to demonstrate the Doppler effect in an experiment.

However, as Doppler himself also realized, there was a way round this problem: like light, sound also travels in waves, only much more slowly than light. Accordingly, the postulated effect would 'also apply absolutely stringently' to sound waves. Sound is a wave that is composed of rapid and small variations in air pressure, which the human ear is able to register. Just like the ship sailing into the waves, sound-wave pulses reach the ear in quicker succession if a person is moving towards the sound source, and this makes the pitch seem higher than that at which it is being emitted by the source. Doppler calculated that a sound source would have to be approaching the hearer at a velocity of 68 ft per second (70 km/h) in order to change a B into a C, a semitone higher.

Seventy kilometres an hour – now, since the invention of the steam locomotive at the end of the preceding century, this was a speed that could be attained. Buys Ballot approached the director of the Dutch Rhine Railway, who in turn got permission from the country's interior minister for the 'free use of a locomotive'.

Buys Ballott's first idea was to use the train's whistle as his sound source. It was loud and therefore audible over a great distance. Yet from preliminary trials he realized that the note it produced was too impure for a musician to be able to determine its precise pitch. And so Buys Ballot expanded his number of assistants by engaging the services of a handful of the best trumpet players he could find in Utrecht. One of them travelled on the railway car with two assistants while the others waited in three groups along the line at an interval of 400 m (1312 ft) from one another.

On the outward journey, in the service of science, the trumpet player on the railway car would play a G, while the musicians beside the track noted the differences in tone. On the return journey, the roles were reversed: now the trumpet players beside the track played, while the musician on the railway car tried to determine the pitch of the note.

Source

Buijs (Buys) Ballot, C. (1845), 'Akustische Versuche auf der Niederländischen Eisenbahn, nebst gelegentlichen Bemerkungen zur Theorie des Hrn. Prof. Doppler' ('Acoustic experiments on the Netherlands railways, together with occasional remarks on the theories of Professor Doppler'). *Poggendorff's Annalen der Physik und Chemie* 66, pp. 321–351.

However simple Buys Ballot thought the experiment might be, its actual execution turned out to be far more tricky. In order to achieve the greatest possible difference in tone, the locomotive had to travel as fast as possible, but the faster it went, the more difficult it became to make out the sound of the trumpets above the noise of the engine. Moreover, at this speed, the train was soon far away, meaning that the note was audible only for a very brief instant. On the other hand, if the train went slowly, then the difference in tone was imperceptibly small. Ultimately, Buys Ballot settled on speeds between 18 and 72 kilometres per hour, which he timed with two watches. To his annoyance, however, the fireman could never manage to keep the speeds constant.

Yet Buys Ballot's main problem was not so much technical as personal: despite being given precise instructions on where to come in, the musicians proved incapable of playing their notes right on cue. Sometimes one of them would forget to play his G, while at others two players would suddenly strike up simultaneously. In *Poggendorff's Annals of Physics and Chemistry*, Buys Ballot advised anyone wanting to repeat his experiment to use 'properly disciplined individuals'.

Once Buys Ballot had repeated the experiments he conducted

with valved trumpets on 3 June with louder natural trumpets on 5 June, he was in a position to confirm Doppler's theory 'despite some irregularities'. The musicians concurred that the note was higher when the trumpet player was approaching than it was when he was moving away from them. Buys Ballot had a ready explanation as to why this effect was not evident in the noise of a passing coach and horses – as some of the musicians had argued before the experiment. A coach didn't produce a pure note, but rather a mixture of various high notes. Detecting any shift in tone from this was impossible, even to a musician's ear.

On similar grounds, Buys Ballot was also convinced that Doppler was mistaken on one point: although his theory was undoubtedly correct, it didn't account for the colour of the stars. The light emitted by stars was also a mixture, and what is more of diverse colours. If, in line with the Doppler effect, all these simultaneously shifted up a notch, then the lowest-frequency light – i.e. red – would actually have been missing from the spectrum.

Doppler believed that this change in colour was visible in binary stars, but overlooked the fact that stars also emit rays in the invisible infrared part of the spectrum. Infrared light waves are slower still than red ones and are quite simply shifted by the Doppler effect into the visible region. Thus, for all practical purposes, where human visual perception of the phenomenon is concerned, absolutely nothing changes. Ironically, Doppler chose to highlight in the title of his paper precisely that phenomenon – the colour of binary stars – that does not come about as a result of the Doppler effect. Stars actually emit coloured light from the outset.

Nowadays, in all likelihood, Doppler wouldn't choose the example of binary stars to corroborate his theory, but rather that of ambulances: every child knows that the siren sounds more high-pitched when the ambulance is approaching and more low-pitched when it's moving away.

Today, countless technical applications in astronomy, chemistry and medicine are based on the Doppler effect. Navigational systems on aircraft rely upon it, the Big Bang Theory could never have been devised without it, and even radar speed traps use it.

Buys Ballot didn't look that far into the future. The only practical application of the Doppler effect that he envisaged was that it 'might one day contribute to the manufacture of better musical instruments'.

madsciencebook.com

An interactive animation on the Doppler effect.

1852 The Lechery Muscle

The physician Guillaume Benjamin Armand Duchenne de Bourgogne never disclosed the name of the old man whose image now appears in countless art exhibitions and is reproduced in many illustrated books. In Duchenne's work *The Mechanism of Human Facial Expression* we learn only that he worked as a cobbler and that his facial features attested to his 'benign character' and his 'limited intelligence'.

Far more than the fate of his experimental subjects, Duchenne was preoccupied with how readers of his book might respond to the fact that he hadn't chosen a more handsome face for his experiments: 'The old man whom I photographed for most of my electro-physical experiments had common, ugly features. This choice might appear odd to a man of the world.'

Guillaume B.A. Duchenne de Bourgogne (right) used electricity to study the 'orthography of facial expression'.

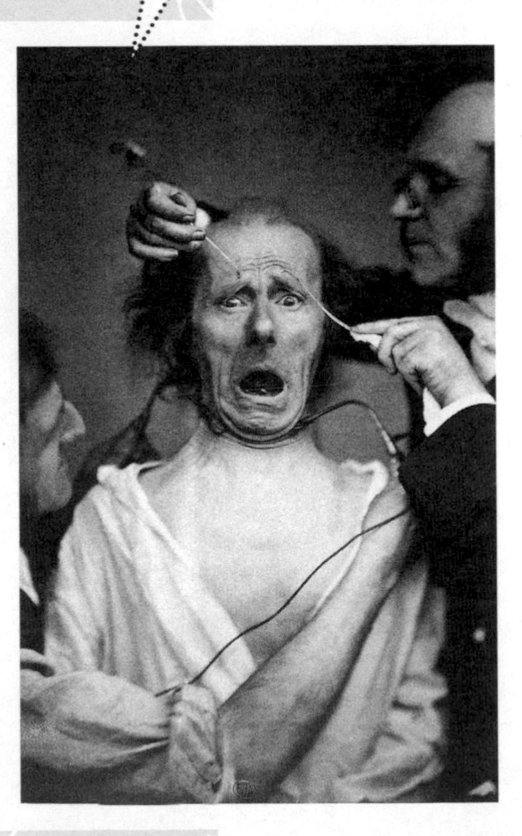

However, Duchenne had good reasons for preferring the toothless dotard. For one thing, his wrinkled skin enabled the musculature of his face to be seen particularly clearly, and for another, he had suffered for a long time from a complete lack of feeling in his face. This was an incalculable advantage that allowed Duchenne 'to examine the actions of individual muscles as efficiently as if they had been on a corpse'. Indeed, he had even taken account of this possibility: 'It's true that I could have used a corpse instead of this living person.' However, Duchenne knew from personal experience that there was no more repellent spectacle than using electricity to produce emotions on the face of a dead person. 'My old man therefore made the ideal experimental subject.' Despite the fact that the images of the cobbler are sometimes reminiscent of torture scenes, he felt nothing and his breathing remained regular and calm throughout the experiments, as the doctor was at pains to reassure his readers.

When the 36-year-old Duchenne moved from Boulogne-sur-Mer on the English Channel to Paris in 1842, he did not take up a permanent position but worked instead as an itinerant physician, treating patients in various hospitals. This included the Salpêtrière Hopital on the banks of the lower reaches of the River Seine, an area inhabited by many patients suffering from a range of paralyses that had never been properly diagnosed. In the course of examining epileptics, spastics and paraplegics, Duchenne stimulated patients' individual muscles with electricity, and in the process built up an extensive catalogue of neurological illnesses.

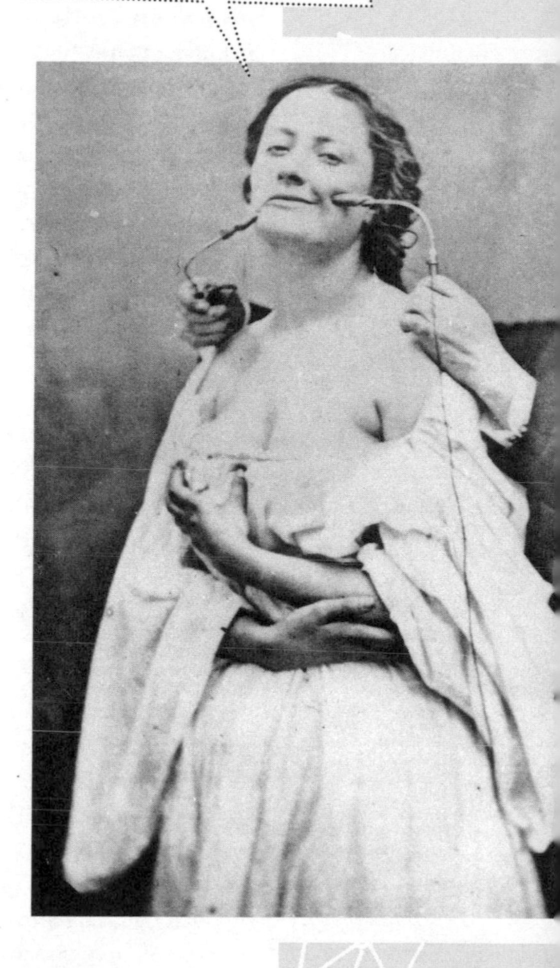

In some cases, Duchenne de Bourgogne preferred to arrange his models in theatrical poses. He entitled this image 'The Seductive Woman'.

Duchenne concluded that, if a paralysed muscle could be stimulated with electricity, then the natural control mechanism must be somehow damaged, and the fault must lie in the brain, or in the connection between the brain and muscle. On the other hand, if it couldn't be thus stimulated, then the problem must reside in the muscle itself. Nowadays, the name of the most common disease associated with the atrophy of muscle tissue – Duchenne muscular dystrophy – recalls his pioneering work in this field.

The muscles that Duchenne investigated included those in the face. In studying these, he wasn't just pursuing scientific aims, but also aesthetic ones. He was convinced that with electrodes and alternating current he could fathom the 'laws that govern human facial expressions' – and in the process unravel the mystery of the universal 'orthography of facial expression' that God had created, and which meant that a particular emotion galvanized the same combination of facial muscles into action in every human being without exception.

In his experiments, Duchenne tried to produce the most authentic-looking emotions by stimulating the facial muscles electrically. Using up to four electrodes at the same time, he was able to create facial expressions signifying rage, delight or surprise; sometimes he also used electricity to evoke a different emotion on either side of the face. He named the muscles after the feelings by which they were activated: the sadness muscle (m. depressor anguli oris), the pain muscle (m. corrugator supercilii), the lechery muscle (part of m. nasalis). He also discovered that the difference between a genuine and a false smile lay in orbicularis oculi, pars lateralis – a muscle that surrounds the eye and that is only activated when a person smiles naturally. This muscle, wrote Duchenne, 'isn't subject to a person's free will … and the absence of such a smile betrays the false friend'.

Source

Duchenne de Boulogne, G.B.A. (1862), *Méchanisme de la physionomie humaine Jules Renouard.* Translation: *The Mechanism of Human Facial Expression* (1990). Cambridge University Press.

Electrostimulation had the disadvantage that the effect of electricity on the muscles lasted only a short time. If photography – which could capture fleeting phenomena – hadn't been invented at precisely that period, then surely the only people nowadays who had heard of Duchenne would be a few neurologists interested in the history of their field. As it stands, the photographs of his experiments have assured him a place in the history of photography as well. Huge sums now change hands for an original plate of the 'ugly cobbler' from Duchenne's first published findings. Charles Darwin also used several of Duchenne's pictures in his 1872 work *The Expression of Emotions in Men and Animals*.

Although the old man was Duchenne's most famous subject, he was by no means his only guinea pig. For example, he also conducted experiments on a young woman whose eye disease he was treating by means of electrostimulation. After she had got used to the uncomfortable procedure, he began to arrange her in theatrical poses: sometimes showing her with an imploring look, at others smiling lasciviously; sometimes as a mother beside the cradle, and on other occasions as Lady Macbeth. There is

"It's true that I could have used a corpse instead of this living person."

something surreal about these photos, given that you can always see Duchenne's hands emerging from the edge of the picture and pressing electrodes to the woman's face.

Duchenne did not regard his work merely as a way of gaining knowledge. He also wanted to use his studies to change the course of art, and formulated rules to guide painters 'in the true and comprehensive depiction of people's emotional states'.

He took a dim view of the efforts of many of the masters of ancient art. Although he thought that they had successfully captured people's basic features, many other aspects of their depiction of human expressions were 'mechanically impossible'. For instance, he thought that the famous Greek sculpture of the priest Laocöon, which art historians praised as a masterpiece, was deficient in its representation of the figure's forehead. Clearly the sculptors Polydoros, Agesandros and Athanodoros from Rhodes knew nothing about the vital subcutaneous effect of m. corrugator supercilii.

madsciencebook.com A test: can you distinguish a genuine smile from a false one?

To show how much more beautiful a person's expression could be when the artist obeyed the 'immutable laws of nature', Duchenne used some plaster to reconstruct how the natural expression should really have looked on a copy of the statue. Nor was this the only piece of classical art that he busied himself with. He rejected the criticism that he was reducing art to the level of 'anatomical realism'; after all, his critique of art was founded on 'strict scientific analysis'.

1883 Great – Someone Else Can Do All the Hard Work!

It had long been common knowledge, but the first person to prove it scientifically was the French agronomist Max Ringelmann at the end of the 19th century: people are lazy. Especially when they think that nobody's watching them.

Ringelmann's elegant experiment consisted of getting 20 students from the Grandjouan Agricultural College to haul on a 16-ft (5-m) rope, the other end of which was attached to a strain gauge. This apparatus confirmed in stark figures the human tendency to shirk. When two people tugged on the rope simultaneously, each of them produced on average just 93 per cent of the effort they had previously expended when on their own. With three people, this figure dropped to 85 per cent, and with four to 77 per cent. From there, the index of laziness continued to spiral downwards, until in a group of eight people, each individual was performing on average at only 50 per cent of their maximum capacity. Modern psychologists refer to this sneaky tendency in human nature as the Ringelmann effect, and explain it in the following terms: in a task involving a group, each person's input does not have as strong an influence on the collective outcome, thus demotivating the individual from giving of their best; furthermore, the individual's contribution is concealed in the group effort, further encouraging freeloading.

A reprise of the 1883 tug-of-war experiment at the University of Massachusetts in 1974. The result was the same: the more people who are pulling, the less each individual pulls his weight.

Yet Ringelmann recognized that his results were also open to another possible interpretation. Could it be that the decline in efficiency had nothing to do with 'social loafing', but rather with the greater difficulty involved in synchronizing effort in a group tug-of-war? So, if it was the case that the students hadn't all pulled together on the rope at precisely the same time, then this would explain the fact that the pull force was smaller than the sum of their individual efforts – and would also restore the reputation of human beings as selfless entities.

Yet this hope was conclusively dashed in the 1970s by Alan G. Ingham from the University of Washington, who devised a modern version of Ringelmann's experiment. Among the real participants, Ingham infiltrated stooges who only pretended to pull on the rope. Each experiment included just one person who wasn't in the know, and he was led to believe that he was working either on his own or in groups of two, three, four, five, six or seven people. So that he didn't get wind that the others weren't pulling their weight, he was always put at the front of the line; alternatively, Ingham would blindfold all the participants on some pretext or other. The upshot was that even the uninitiated individual's motivation to do his best decreased; what's more, he scaled down the effort he expended irrespective of the number of supposed colleagues he was working with.

Since the concept of teamwork became firmly established in the modern workplace, Ringelmann's study sometimes crops up in management courses. But there's really no point in them going into it in depth: Ringelmann's inescapable conclusion has long been summed up in the old quip about the true definition of the word 'team': Great – someone else can do all the hard work.

Source

Ringelmann, M. (1913), 'Recherches sur les moteurs animés: travails de l'homme' ('Research on animate sources of power: the work of man'). *Annales de l'Institut National Agronomique* 2 (12), pp. 2–39.

Ingham, A.G., et al. (1974), 'The Ringelmann Effect: Studies of Group Size and Group Performance'. *Journal of Experimental Social Psychology* 10, pp. 371–384.

1885 The Murderer's Head

The French physician Jean Baptiste Vincent Laborde was relieved to find that people in the countryside weren't so hung up about the letter of the law as they were in Paris. His work in the capital was being hampered by a 'stupid law' that 'throughout the whole of civilized Europe exists only here'. This statute required that the body of an executed person be transported to the gates of the cemetery and given a sham burial there 'instead of immediately releasing it in a condition where it could be used for scientific examination', as Laborde bitterly complained.

Laborde's aim was to find out how long a person's head could live after being severed from the torso. Since the introduction of the guillotine in France in 1791 as a supposedly just and humane method of execution, the question had arisen of how humane it really was. Some medical experts claimed that consciousness and the capacity to feel pain were still present even a quarter of an hour after the person had been beheaded. The matter of the precise moment of death was also a hot topic for literature. In Victor Hugo's short story 'Le Dernier Jour d'un condamné' ('A Condemned Man's Final Day'), the death-row prisoner writes in his diary: 'Once the deed is done, so people say, there's an end to all suffering, but how can they be certain? Who told them so? Who ever heard of a severed head balancing on the rim of the basket and calling out to the assembled crowd: "It didn't hurt a bit!"?' In a novel, *Le Secret de l'échafaud* ('The Secret of the Scaffold') by Villiers de L'Isle-Adam, the surgeon Armand Velpeau tries to dispel all uncertainty on the matter by making a compact with the condemned doctor Edmond-Désiré Couty de la Pommerais: after being beheaded, the doctor is to respond to a prearranged signal and wink three times – if he is indeed still conscious at this stage.

These scenes weren't just the figment of novelists' imaginations. Using some highly inventive methods, scientists had long attempted to answer the question of the exact moment of death. In some cases, severed heads were punched, while in others, researchers yelled at them or shouted their name and waited for a reaction. Laborde's technique was a little more original. He had already tried several times to connect severed human heads to the circulatory system of a living dog, but the precious minutes that were lost thanks to the 'stupid law' were a key factor. So as to waste not a moment, he once waited at the gates of the cemetery for a hearse to deliver the body of a beheaded criminal and started to operate on the still-warm head during the bumpy coach ride back to his lab.

Out in the sticks, things were far simpler. And so it was that Laborde found himself standing on the Place de la Tour in the small town of Troyes, 93 miles (150 km) east of Paris on 2 July 1885, eagerly awaiting the execution of a murderer by the name of Gagny. Six months earlier, at the Gloire-Dieu farmstead, Gagny, together with an accomplice, had murdered the owner, his mother and their maid.

With the backing of a doctor from Troyes and the approval of the town's mayor, Laborde took possession of Gagny's head seven minutes after his execution and immediately set about connecting his left carotid artery to that of a large dog. Through the right carotid of the murderer's head, he planned to introduce warmed ox blood with a syringe. But he was frustrated in this when it became apparent that guillotines in the country were not so well maintained as those in the city. 'The cut had been poorly executed and the victim's tissue was all squished and lacerated, and this made it extremely difficult to locate the carotid arteries,' Laborde later noted. However, even without maintaining a flow of blood, holding a candle up in front of the head's eyes still produced an effect: the pupils narrowed. Finally, after 20 minutes, the double transfusion was able to proceed.

The effect was instantly noticeable: 'The left side especially, which was being supplied with the dog's blood, took on a purplish hue, which surprised those who had not witnessed my earlier experiments.' Through holes that he had drilled in the skull, Laborde now began to apply electric shocks to the brain. But even when he applied the maximum charge, nothing happened. 'As time passed, the faces of many people began to register disappointment.' Undaunted, Laborde drilled some new holes. And it was through one of these, on the right-hand side, that he struck lucky: the electrical stimulation of the brain at this point provoked muscular spasms on the left side of the face. Forty minutes after the execution, the teeth in the head were even heard to chatter, as Laborde proudly recorded.

The knowledge yielded by these macabre experiments didn't ultimately amount to much. According to Laborde, they showed that the brain remains active after death for twice as long if blood is administered than if it isn't. However, Laborde never found out what implications this might have for an executed person – namely, whether a severed head still retains consciousness and, if so, for how long.

Source

Laborde, M. (1885), 'L'exitabilité cérébrale après décapitation (1). Nouvelles recherches sur deux suppliciés: Gagny et Heurtevent' ('Brain activity after decapitation: new researches on two execution victims, Gagny and Heurtevent'). *Revue scientifique 2e semestre*, pp. 673–677.

1889 Guinea Pig Testicles – the Secret of Youth

Charles-Édouard Brown-Séquard wasn't remotely squeamish when it came to risking his own life and limb in the cause of medicine. This eccentric doctor had already injected his own blood into the bodies of

beheaded men, eaten vomit brought up by cholera patients, and swallowed sponges tied on strings which, once they were saturated with gastric juices, he then spat out.

Yet none of his experiments had such far-reaching consequences as the one he embarked upon on Wednesday 15 May 1889. It was on that day that he ground the testicles from a healthy puppy in a pestle and mortar in his laboratory at the Collège de France in Paris, added a drop or two of distilled water, filtered the resulting suspension, and then injected the liquid into his left forearm.

Brown-Séquard thought that 'infirmity in old age may in part be attributed to deteriorating functioning of the testicles'. The symptoms of frailty associated with aging were identical with those exhibited by eunuchs from early childhood on. And similar disorders were also thought to affect men who masturbated too often. This seemed to the physician to point to a compelling conclusion – namely, that the testicles must secrete a substance into the blood that had an energizing effect throughout the entire body.

The eminent physician Charles-Édouard Brown-Séquard thought that injections of a solution made from ground animal testicles would rejuvenate him.

By the same token, Brown-Séquard was also certain that something could be done to combat aging – even his own advancing years. He was 72 years old, was often forced to take rests from his work in the laboratory, and suffered from insomnia and constipation. The exotic brew would, he hoped, 'halt or at least retard changes in tissue structure that are associated with aging'. He repeated the injection on the following two days, and when his supply of dog testicle extract ran out, he switched over to the miniature testicles of guinea pigs for his next four injections.

Brown-Séquard fancied that he could feel the injections working even on the second day of the experiment. He found that he could dash up flights of stairs once more, stand for long spells at the lab bench, and work late into the night on articles. The treatment also left its mark on the way he passed water, 'specifically

A packet of the rejuvenation compound. Brown-Séquard supplied it to doctors free of charge so long as they agreed to let him study the case histories of the patients they treated with it.

LABORATOIRE DE MEDECINE DU COLLÈGE DE FRANCE
12, Rue Claude-Bernard. — PARIS
EXTRAIT ORGANIQUE
pour expériences scientifiques
Envoi gratuit de MM. BROWN-SEQUARD et d'ARSONVAL
N B. — Sous aucun prétexte ce produit ne peut être vendu.

regarding the distance that my stream of urine covers before hitting the ground in the pissoir': in surely one of the strangest measurements he ever took, he recorded an improvement of at least 25 per cent in his performance in this respect. On 1 June 1889 he announced his results at a meeting of the Biological Society in Paris. His address contained one vague but ultimately telling sentence that really captured the public's imagination: 'I might add that other powers, which admittedly hadn't deserted me entirely but were incontestably debilitated, have also improved markedly.'

In no time, the newspapers were full of talk of an 'elixir of life', and quacks began offering treatment with the new wonder drug. Such treatments resulted not in a longer life but in a severe case of blood poisoning. Brown-Séquard himself didn't earn a penny from the testicle extract; rather, he let doctors have his 'Extrait Organique' for free in return for the case histories of patients treated with it. However, he couldn't prevent his name from being misused to promote all kinds of dubious medication. One of these, called 'Sequarine', claimed that it contained 'essence of animal energy' and that it was effective against everything from anaemia to flu.

The general view nowadays is that the signs of rejuvenation that Brown-Séquard detected in himself were merely the result of the placebo effect. In any event, they were impossible to reproduce. Yet his crude experiment was a precursor of hormone therapy, which today is one of the most commonplace medical procedures. Brown-Séquard didn't live to see

Source

Brown-Séquard, C. (1889), 'Des effets produits chez l'homme par injections sous-cutanées d'un liquide retiré des testicules frais de cobaye et de chien'. ('Effects produced in humans by subcutaneous injections of fluids extracted from the testicles of guinea pigs and dogs.') *Comptes Rendus de la Société de Biologie* 41, pp. 415–422.

the birth of this field of medicine. He died on 2 April 1894 in Paris at the age of 76, barely five years after the experiment.

Ridicule from those who opposed his methods followed Brown-Séquard to the grave, including a rumour that he had died in London on the eve of giving a lecture entitled 'How I Grew 20 Years Younger'.

1894 Dog Tired

You couldn't devise a more drastic illustration of the importance of sleep. The Russian scientist Maria de Manacéïne kept four puppies awake so long that they died. The first expired after 96 hours, and the last one after 143. She attempted to revive six further dogs after keeping them awake for periods of between 96 and 120 hours. All to no avail – they died as well. According to the researcher, the experiment demonstrated 'that total sleep deprivation is more lethal for an animal than a complete lack of sustenance'. Dogs were capable of recovering all on their own even after enduring 20–25 days without food.

This work conclusively rebutted the unorthodox view taken by some scientists that sleep was just a useless habit. Yet de Manacéïne never discovered what it was exactly that the dogs died of. To this day, nobody really knows why the higher animals need sleep.

The main reason cited by de Manacéïne for not continuing with these experiments was that she found them 'extremely arduous'. She omitted to mention how arduous they were for the dogs.

Source

de Manacéïne, M. (1894), 'Quelques observations experimentales sur l'influence de l'insomnie absolue' ('Some experimental observations on the effect of total insomnia'). *Archives italiennes de biologie* 21, pp. 322–325.

1894 Low-flying Cats

In 1894 the Paris Academy of Sciences issued a general appeal for anyone to 'explain how, according to the laws of physics, a cat will always manage to hit the ground paws first after falling from a great height'. To the layman, this presented no problem whatever: the cat simply manoeuvred itself so adroitly in flight that its paws were facing downwards when it landed. By contrast, anyone who knew a thing or two about physics would have concluded that the cat

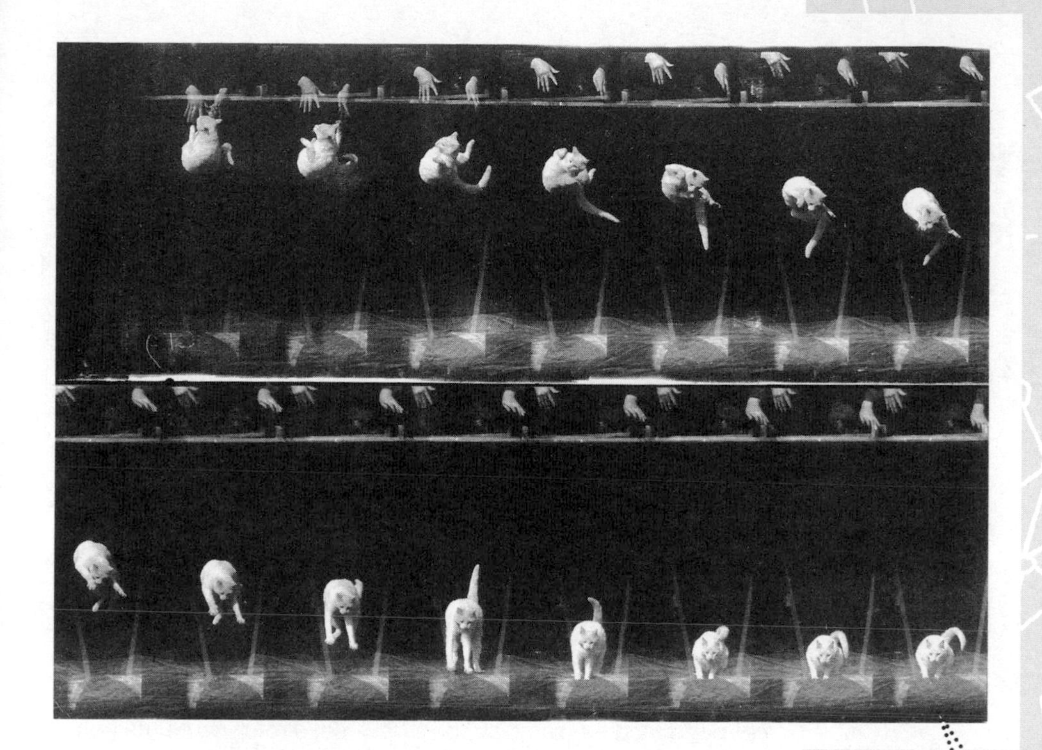

How does a cat always manage to land on its paws? The physician and inventor Étienne-Jules Marey finally solved the mystery with this sequential photograph.

was performing an impossible feat.

The problem was that the falling cat had no surfaces off which to push itself. Thus, every rotation that it made with the front part of its body would, of necessity, cause a rotation in the opposite direction in its hindquarters. For instance, a half-twist to the right up front would create a half-twist to the left behind. So by rights the cat should actually land with its body twisted, which patently wasn't the case.

At the outset, scientists still clung to the view that the cats must be pushing themselves off from the researchers' hands. Yet even when they suspended one of the animals by individual cords attached to each of its paws before releasing it, it still managed to rotate itself. Likewise, the hypothesis that it was somehow using the air to right itself also proved untenable.

The puzzle was finally solved by a physiologist, Étienne-Jules Marey. Marey was an obsessive inventor who constructed a vast array of mechanical devices. Among these was a film camera that could capture the falling cat at a speed of 60 frames per second. Even

Source

Marey, E.-J. (1894), 'Mécanique animale: Des mouvements que certains animaux exécutent pour retomber sur leurs pieds lorsqu'ils sont précipités d'un lieu élevé' ('Animal mechanism: the movements that certain animals make to land on their feet after being dropped from an elevated position'). *La Nature*, pp. 369–370.

after the first showing of the film, some of the physicists present still questioned whether the rotation was possible unless the cat had some surface against which to gain purchase. However, one of them realized how the cat was pulling off its trick.

Marey's original footage of falling cats, dogs and rabbits.

madsciencebook.com

The animal's movement occurred in two stages: first, the cat rotated its forequarters to face the ground before twisting its hindquarters into the same attitude – in the same direction. Between these two movements, an adjustment in the position of the cat's legs enabled the two parts of its body to push off from one another. The cat was employing the same principle as that of an ice-skater performing a pirouette; the skater spins fast when she lays her arms flat against her body and slowly when she holds them out. The cat did both simultaneously, pulling its forepaws into its body and stretching its hind legs away from its body. In this way, it was quickly able to rotate its forequarters with a half-turn to face the ground, to which movement the hindquarters – as a result of the extended back legs – offered resistance and themselves only rotated slightly in the opposite direction. In order to now manoeuvre the hindquarters into a landing position, the cat performed the same action in reverse, stretching out its front paws and tucking its back paws into its body.

Marey's sequential images of movement created a vogue for filming falling animals. Soon dogs, rabbits, apes – and in one experiment a 'small, plump guinea pig', which to the astonishment of the researchers effortlessly twisted its belly through 180 degrees – were all being dropped. The scientists blindfolded their subjects and also conducted tests with animals that had no tails or which were missing their organs of equilibrium. Even a cat with both of these missing managed to rotate itself with no problem. Evidently, cats mainly used their eyes to orientate themselves.

In the 1960s a researcher summed up 70 years of research into falling cats: 'It turns out that the twisting cat raises a number of interesting problems, whose ultimate solution, however, is probably of no great practical value – except, that is, to other cats.'

1895 Sleepless in Iowa

At first glance, this experiment appears quite harmless: three men stayed awake for 90 hours so that the scientists G.T.W. Patrick and J. Allen Gilbert

from the psychological laboratory at the University of Iowa could study the effects of sleep deprivation. But why 90 hours in particular? In their paper on the experiment, Patrick and Gilbert provide no answer, but it seems reasonable to suppose that they got this figure from the work of Maria de Manacéïne. A short time before, this Russian scientist had conducted sleep-deprivation experiments on dogs (see p. 46); all the animals she experimented on died, the first of them after 96 hours.

It's unclear whether any of the participants in Patrick and Gilbert's experiment knew about this. In any event, the first test subject, who appears in the report on the experiment under the initials J.A.G, got up at 6 o'clock on the morning of Wednesday 27 November 1895 and only went to bed again three days later, at midnight on the Saturday. During daylight hours, he went about his normal occupation, and at night he whiled away the time by playing games, reading and taking walks. During the final 50 hours of the experiment, he had to be placed under constant surveillance, since he would otherwise have dropped off at the first opportunity. After the second night of wakefulness, hallucinations began to set in. The test subject complained that the floor was covered with a 'greasy looking, molecular layer of rapidly moving or oscillating particles' that prevented him from walking properly.

Every six hours, J.A.G. and the other participants had to complete a two-hour test. As they were increasingly deprived of sleep, there was a noticeable decline in their ability to concentrate and to remember things.

Patrick and Gilbert were also keen to know how deeply the subjects slept after their long period of wakefulness. To this end, they subjected one of the participants, who was sleeping alone in a room, to electric shocks of increasing intensity after every full hour of sleep. He had been given instructions to press a button next to his bed whenever the shocks woke him.

Yet it turned out that the most powerful electric shock that the apparatus could deliver on its own still wasn't enough to wake the subject. This was only achieved by delivering more intense shocks manually. The deepest period of sleep for the subject came after two hours; then, it was impossible to rouse him sufficiently to press the button. Instead, he would just react to the shock by yelling in pain.

Source

Patrick, G., and Gilbert, J. (1896), 'On the Effects of Loss of Sleep'. *The Psychological Review* 3 (5), pp. 469–483.

1896 The World Turned Upside Down

It was 12 o'clock midday when George Stratton set his brain a task that nobody had ever been faced with before. The 31-year-old psychologist from the University of California in Berkeley strapped a kind of mask to his face, comprising a padded plaster cast of the area around his eyes. On this mask, in front of his right eye, a short tube containing four lenses poked out. His left eye was covered with a blank piece of plaster. This contraption immediately made every action of his into a laborious sequence of short movements and constant corrections: the lenses covering Stratton's right eye turned his world upside down. What had formerly been at the top of his field of vision was now at the bottom, and vice versa. His aim was to keep this mask on for seven days in order to observe how his brain coped with his new perspective on the world.

Stratton's experiment was intended to solve a centuries-old puzzle. In 1604 the astronomer Johannes Kepler had described how an image is formed on the human retina. A couple of years later, someone scraped the membrane covering from the back of an ox's eye and confirmed that Kepler was correct: the light rays crossed over in the lens of the eye. The image that strikes the retina shows the world turned on its head.

How is it, then, that we still see everything the right way up? Although it's quite natural to pose this question, it's also completely pointless. There's no little man in our brains who spots the inverted image on a screen and notices that it's the wrong way round. The network of brain cells that process the signals coming from the eye does not recognize any concept of 'up' or 'down'. The brain merely forms a unified impression from images, sounds and taste sensations so as to ensure, for example, that we feel our foot to be in the position where we see it, and vice versa.

Yet a second question arises: is the inverted image on the retina a necessary condition of our being able to see things in an upright position? Or could the brain also get used to a different orientation?

At the start of the experiment, Stratton experienced a slight feeling of nausea. Every time he turned his head, everything around him seemed to swim about. The orthodox perspective on the world proved tenacious: whenever Stratton called to mind an object that he had just seen upside down, his brain immediately turned it the right way up again. If he reached out to grasp something, he invariably moved the wrong hand. He made notes without looking down at the paper, since the unaccustomed way of looking at things would have made writing impossible. And yet the longer the experiment went on, the more his brain got used to the changed circumstances. By the fifth day, Stratton had regained the ability to walk through the house without having to feel his way around everything.

Readjustment took longest where the perception of his own body was concerned. Stratton's brain was constantly trying to create a logical whole from the contradictory signals coming from his eyes, ears and skin. As long as he couldn't see his arms and legs, he felt them to be in their familiar locations, but as soon as they came within his field of vision and he punched himself, his brain registered the blow as having come from where he now perceived the leg to be. This gave rise to some bizarre hallucinations: if Stratton could see only one of his feet, he could not help seeing it as the other foot in its familiar configuration, i.e. rotated through 180 degrees and facing in the opposite direction.

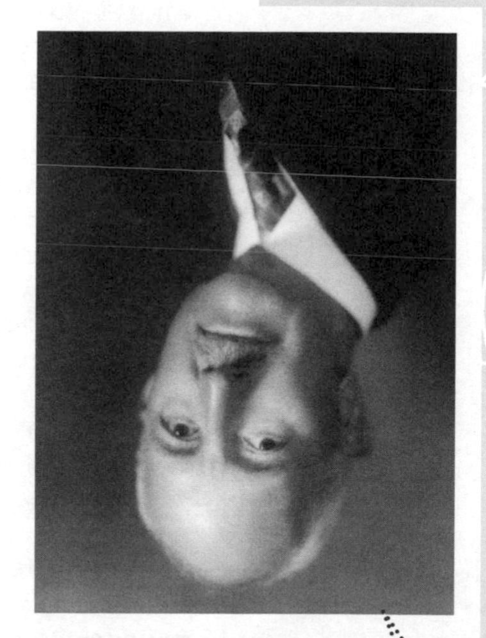

On your retina, this image of the psychologist George Stratton is the right way up. So why does it appear upside down here?

What's more, his sense of sight always took precedence over his hearing: compared to when he had his old form of perception, his footsteps seemed to come from the opposite direction. Only those parts of his body that he could not see in the limited field of vision afforded by his inverting glasses resisted this process of reversal. Although, when eating, his new sense of sight told him that he was moving his fork up to a place above his eye, the illusion that that was where his mouth was now located was instantly dispelled by the food touching his lips. He occasionally managed the Cubist feat of feeling his forehead to be below his eyes, and, for an instant, his mouth was located on his brow.

Textbooks often give the impression that, towards the end of the experiment,

Source

Stratton, G.M. (1897), 'Vision without Inversion of the Retinal Image'. *Psychological Review* 4, pp. 341–360, 463–481.

Stratton had long periods when he could perceive the world the right way up again. In actual fact, he was able to conjure up an image of an upright world only by concentrating very hard, and even then it only appeared fleetingly.

Nevertheless, after 87 hours wearing the inverting glasses – at night, he wore a blindfold – Stratton came to the conclusion that 'The inverted image on the retina is not, therefore, essential to "upright vision".' In other words, when faced with a skewed image, the brain was capable of restoring harmony between what a person perceives and what they feel.

This harmonizing of the senses holds the key to what is really meant by 'upright vision'. For the perceived object cannot in and of itself be either upright or inverted, but can only be so in relation to what the other senses tell us. The fact that Stratton's world, even towards the end of the experiment, was still mostly upside down had nothing to do with his brain not having come to terms with his new perception, but rather with the fact that it could still remember what the world had looked like before.

When Stratton finally took off the inverting glasses, although what he saw seemed alien to him, it was the right way up. He used the wrong hand to reach out for things and ducked down when he should have been stretching up. However, after a day these aberrations disappeared.

Stratton's experiment was repeated and extended. For instance, the test subjects wore mirrors that deflected their vision in such a way that they felt they had eyes in the back of their head. The results were basically the same. The brain can adapt to such an extent that test subjects wearing inverting glasses can even go mountain climbing or ride a bike in rush-hour traffic.

1899 Bodies in the Vegetable Patch

Between May 1899 and September 1900, people of a nervous disposition would have been well advised to avoid taking a constitutional in the grounds of Krakow University's Medico-Legal Institute. For, during this period, the pathologist Eduard Ritter von Niezabitowski was conducting his experiments on the 'theory of cadaver fauna'. This entailed exposing the bodies of stillborn children 'to the actions of corpse-eating insects in a little-visited area of the main vegetable garden surrounding the institute buildings'. For the purposes of comparison, he buried the bodies of calves, cats, rats and moles there as well.

Niezabitowski wanted to find out various things: the sequence in which carrion insects colonize a dead body; whether the insects that devour a human corpse differ from those that land on animal corpses; what influence the season has on cadaver fauna; and how long it takes before a body is reduced to just the skeleton. He visited the garden every day and collected the insects feasting on the decomposing bodies, and found on one corpse no fewer than 11 different species. However, it was evident from the very first day that the greatest amount of material – some three-quarters of the bodies' soft tissues – was being eaten by maggots of the green bottle fly Lucilia caesar l. Another hard worker was the beetle Necrodes litoralis, which got down to business only when the body had been dead around a week. In summer it took a fortnight for the corpse to be reduced to a skeleton, and somewhat longer in spring and autumn. Niezabitowski was unable, however, to confirm the claim that had sometimes been made that a quite unique mixture of insects feasts on the human body. He found the same species on the animal cadavers.

Knowledge about the sequence in which insects colonize a corpse can be used to draw conclusions about the time of death, Niezabitowski wrote, but these findings would only be valid for the specific locality where the body was found. Nowadays, forensic entomology is a well-established branch of criminal investigation.

Source

Niezabitowski, E.R. von (1902), 'Experimentelle Beiträge zur Lehre von der Leichenfauna' ('Experimental contributions to the theory of cadaver fauna'). *Vierteljahresschrift für gerichtliche Medizin und öffentliches Sanitätswesen* (1), pp. 44–50.

1899 Tearing out Pubic Hairs

Perhaps it was nervousness, or perhaps the wrong needle had been laid out ready for use. But when August Hildebrandt started to inject his boss Albert Bier with the syringe, everything began to go badly wrong: far too much spinal-marrow fluid drained out of the entry hole and most of the cocaine solution missed its mark. The time was around 7 o'clock on the evening of 24 August 1898, and what then ensued was as much a groundbreaking experiment as it was a black comedy. It was to end with Bier becoming a star doctor and Hildebrandt a research assistant with a fine collection of bruises, puncture wounds and burns.

August Bier was the chief physician at the Royal Surgical Hospital at the University of Kiel and had already tried out a new method of anaesthesia in a number of operations to amputate legs; this involved injecting a solution of cocaine into the patient's vertebral canal, where all the body's nerves converge, arranged according to the parts they relate to. For instance, those controlling the arms, shoulders and chest are found in the upper part of the canal, while the lower contains those governing the lower body and legs. The cocaine solution had a numbing effect on these nerves, so that, depending on the position and the strength of the injection, the patient's body could be rendered insensitive to pain to a certain degree.

Although laughing gas, ether and chloroform were already being used in those days as anaesthetics, they tended to put the patient into a state of deep unconsciousness, which could have fatal consequences if the dosage was wrong.

The physician August Bier tried out a new method of anaesthesia – on his assistant.

Spinal anaesthesia offered a way out of this problem, and after having used it on his patients, Bier was keen to test it out on himself. However, as a result of the mishap with the needle, he had lost a great deal of spinal-marrow fluid and was minded to postpone the experiment. But then his assistant Hildebrandt put himself forward as a test subject. At around 7.30 p.m., Bier injected him with half a cubic centimetre of a 1 per cent solution of cocaine. He then began to transcribe the proceedings:

'After 10 minutes' a large, curved surgical needle was jabbed into the subject's leg right down to the thighbone without causing the slightest sensation of pain.

After 13 minutes, a lit cigar pressed against the legs was registered as being hot but not painful.

After 20 minutes, when the subject's pubic hair was torn out, it just felt as though a fold of skin was being lifted, whereas plucking hairs from the chest above the nipples was acutely painful. The subject reported that the sensation of having his toes bent back too far was not unpleasant.

After 23 minutes, a sharp blow with a lump hammer against the shinbone caused no pain.

After 25 minutes, strong pressure and tugging on the subject's testicles caused no sensation.'

After three-quarters of an hour, normal sensation returned, and the two men went for a meal, at which they drank wine and smoked several cigars. As Bier later wrote, they 'overindulged'. He was laid up in bed for nine days thereafter with a headache. Hildebrandt fared even worse, becoming nauseous and suffering unbearable headaches and severe contusions; all the while, his whole body was racked with pain.

After the results of the experiment were published, spinal anaesthesia quickly became widespread. It is now a perfectly routine medical procedure (albeit no longer using cocaine).

Hildebrandt later turned against his former boss, claiming that it was not Bier but the American James Leonhard Corning who was the father of spinal anaesthesia. While it is true that Corning had conducted similar experiments, it was Bier who recognized their potential.

To this day, nobody knows why Hildebrandt turned on Bier. Perhaps his difficult personality was to blame: Hildebrandt was widely regarded as being stand-offish and having a nasty temper. Alternatively, might it have been the fact that Hildebrandt was mentioned in Bier's article merely as a test subject and not as a co-author? And so it was that he was recorded for posterity in medical history as the assistant who had his balls grabbed by his boss.

Source

Bier, A. (1899), 'Versuche über Cocainisirung des Rückenmarkes' ('Experiments on the cocainization of the spinal marrow'). *Deutsche Zeitschrift für Chirurgie* 51, pp. 361–369.

1900 Rat on a Detour

In 1690, to the southwest of London, the royal gardeners George London and Henry Wise began laying out a maze. At the behest of King William III, they planted small beech trees to mark out 2625 ft

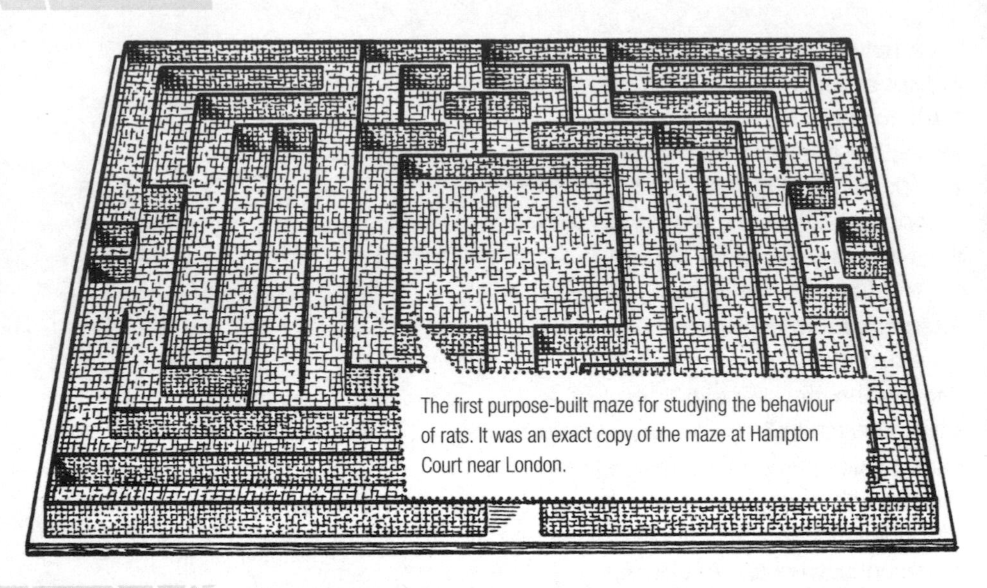

The first purpose-built maze for studying the behaviour of rats. It was an exact copy of the maze at Hampton Court near London.

(800 m) of winding paths in the grounds of Hampton Court Palace – paths on which some 330,000 visitors still lose their way every year.

Two hundred years later, the psychologist Willard S. Small used some chicken wire and a wooden board to construct his own Hampton Court Maze – for his lab rats.

This researcher from Clark University in Worcester, Massachusetts, got the idea when he was searching for a way of finding out how intelligent rats are. The experiment had to be conducted in a controlled environment, but as far as possible the rats' behaviour was not to be affected by unnatural conditions.

Since rats have a predilection for winding passageways, Small lighted on the idea of building a miniature labyrinth. In all likelihood, he was influenced by an article he had recently read about kangaroo rats. An illustration in this piece that showed their burrows 'bore a striking resemblance to the apparatus used in these experiments', as Small later wrote.

He couldn't have imagined how far-reaching a resonance the contraption he devised for his experiment would

'They seem to be getting used to their new surroundings.'

have, not just in psychology but in the wider world as well. The rat in the maze came to epitomize scientific research per se, but also became a metaphor for a person who is unable to get his bearings in the labyrinth that is modern life. If you type the term 'like a rat in a maze' into an internet search engine, you get literally thousands of hits. For example, someone out there is claiming that the US government is acting 'like a rat in a maze', or a certain Chris says on his website that he feels like one whenever he wakes up on a Monday morning. The site www.achievinghappiness.com provides a so-called 'Happiness Formula' that is designed to stop you from feeling 'like a rat in a maze'. The image has also spawned an entire cartoon genre – no other piece of apparatus used in scientific experiments appears more often in cartoons than the maze.

'If you ask me, you don't look much like an experimental psychologist.'

'I don't normally volunteer for experiments, but I'm a real fan of puzzles.'

The reason for Small copying the Hampton Court Maze in particular had to do with the fact that he looked up the term 'Labyrinth' in the Encyclopaedia Britannica and chanced upon a map of the famous English maze. His version was a square measuring 8 ft x 6 ft (2.4 m x 1.8 m) – the Hampton Court Maze is trapezoid in shape. He separated the passageways off from one another with pieces of chicken wire 4 in (10 cm) high. He strewed sawdust on the floor of the maze and put some food in the centre.

Then he introduced his rats to the labyrinth. The first two failed the test because noise in the laboratory spooked them. The third rat, a male, found its way to the centre after 15 minutes, while the fourth took 10 minutes, the fifth 1 minute 45 seconds, the sixth

Source

Small, W.S. (1900–1901), 'Experimental study of the mental processes of the rat'. *American Journal of Psychology* 12, pp. 206–239.

3 minutes, and the seventh 50 seconds. Each time a rat was reintroduced to the maze, it knew its way around better.

Though this might seem a somewhat banal result, it was actually quite astonishing, for it was only when the rat located the food that it learned which of the many routes it had tried to take was the correct one. In other words, it was clear that it had the capacity to memorize where it had turned left or right five minutes previously.

On the basis of these experiments, the psychologist Edward Lee Thorndike formulated his 'law of effect', which states that actions (such as discovering the correct route through a maze) that have satisfying consequences for an organism (namely, finding food) are more likely to recur than those that lead to unpleasant consequences. Thirty years later, the psychologist B.F. Skinner was to apply a new set of terms to this basic theory, invent another piece of apparatus that became a favourite with cartoonists, and earn himself much opprobrium with his hypotheses concerning human behaviour (see p. 97).

"Like a rat in a maze"

1901 Attempted Murder in the Lecture Theatre

As planned, the gun goes off at a quarter to eight. The date is 4 December 1901, and Professor Franz von Liszt has just finished delivering his lecture on the theories of the French jurist Gabriel Tarde at Berlin University's department of criminology. Then, suddenly, a member of the audience stands up and launches into a speech.

'I'd just like to consider Tarde's theories for a moment from the standpoint of Christian moral philosophy.'

'That's all we need!' interposes the person sitting next to him. An angry exchange ensues.

'I'd be obliged if you kept your opinions to yourself. Nobody was asking what you thought.'

'How dare you!'

'One more word out of you, and ...' The first speaker

madsciencebook.com Are you a reliable eyewitness? Take the test and find out.

raises his fist, threatening his neighbour – whereupon the other man draws a revolver and presses the muzzle up against his opponent's temple.

Professor von Liszt rushes across and tries to wrestle down the arm of the man holding the revolver. Just as the gun is pointing at his heart, it goes off.

Those attending the lecture could not know that the weapon was just a toy gun and that the macabre piece of theatre they had just witnessed was part of an experiment devised by the German psychologist William Stern. Stern was a jack-of-all-trades who dabbled in all areas of psychology. He was the progenitor of the concept of IQ, was involved in developmental psychology, and was the publisher of *Beiträge zur Psychologie der Aussage* ('Contributions to the Psychology of Witness Testimony'). In this professional journal, researchers grappled with the question of the degree of exactitude with which people recall events.

Getting test subjects to describe a picture that they had just spent 45 seconds looking at, Stern had noticed that most people's memories are far from flawless. Many swore blind that they had seen certain things in the picture that simply weren't there. The question of the reliability of memory was especially crucial in legal proceedings. And so Stern suggested the experiment with the staged argument, where witnesses were presented with a situation closely resembling a real crime.

As soon as the revolver was fired, those present instantly became aware that the argument was all a sham. Fifteen of the audience, who were either 'more mature students of law or pupil advocates', were subsequently required to provide written or oral witness statements. Three of them were tested either on the same evening or the day after, nine a week later, while the final three were questioned only after five weeks had elapsed since the incident. Not one of them could remember all the details of the incident, which had been split up into 15 individual stages. The rate of error was somewhere between 27 and 80 per cent.

As anticipated, many of the witnesses could not exactly recall what had been said during the incident. But what was really surprising was that some witnesses also fabricated events that had never taken place. For instance, they put words in the mouths of onlookers who had said nothing, and had one of the men involved in the argument fleeing the scene before the other, although in actual fact both had stayed put throughout.

Source

Jaffa, S. (1903), 'Ein psychologisches Experiment im kriminalistischen Seminar der Universität Berlin' ('A psychological experiment in the criminology faculty at the University of Berlin'). *Beiträge zur Psychologie der Aussage* (1), pp. 79–99.

The lack of reliability of witness testimonies provoked a lively debate among lawyers. As Franz von Liszt himself put it in the *Deutsche Juristen-Zeitung* ('German Legal Journal'), 'What, then, is to become of our entire criminal justice system if its surest foundation – namely, the testimony of unimpeachable eyewitnesses – is placed in serious doubt by rigorous scientific research, and if our faith in the reliability of our most valuable source of evidence is undermined?' William Stern, who had instigated the experiment, argued in favour of calling experts to the witness stand to help the court assess the reliability of sworn statements. Indeed, this practice is now commonplace.

The method employed in the revolver experiment – the so-called 'surprise test', where the subjects have no idea that they are taking part in an experiment – became fashionable in the early years of the 20th century. In one case, students were quizzed about a loud, phoney argument that broke out outside the door to the lecture theatre, and in another about a mysterious masked figure who had sat in on the lecture for 20 minutes. In the latter instance, only 4 of the 22 people present were able to pick the mask out from among nine other masks a few days after the lecture.

Occasionally, a sense of theatricality would get the better of those staging these experiments. In 1903, for example, in the middle of a lecture being given to the Göttingen Society for Psychiatry and Forensics, 'a clown clutching a pig's bladder in one hand and a red fez in the other' burst into the room, closely followed by 'a negro in a flamboyant outfit holding a revolver'. The audience were subsequently asked to fill out a questionnaire on the incident, in which they got things completely confused.

The finding that people's memories were intrinsically unreliable was confirmed most succinctly in the reporting of the actual revolver experiment itself. In a textbook on forensic psychology published in 1955, the gun attack in Berlin had transmogrified into a simulated 'deadly assault with a knife'.

1901 The Soul Weighs 21 Grams

The announcement was so momentous that even the *New York Times* reported it, its edition of 11 March 1907 carrying the headline 'Soul Has Weight, Physician Thinks' on page 5. The story described

the curious experiments being conducted by one Duncan MacDougall, a medical practitioner from the town of Haverhill, Massachusetts.

MacDougall had long been interested in the nature of the soul. According to his perverse reasoning, if psychic functions continue after a person's death, then they must have occupied a certain amount of space within the living body. And since everything that takes up space must 'according to the latest conception of science' also weigh something, then it should be possible to determine the weight of the soul 'by weighing a human being in the act of death'. And so MacDougall constructed a precision platform-beam weighing machine, which consisted of a bed suspended from a frame, whose weight – including that of the person lying in it – could be measured to within 5 grams.

However, the sheer sensitivity of these scales severely restricted the choice of test subject. As MacDougall later reported in the professional journal *American Medicine*, 'It seemed to me best to select a patient dying with a disease that produces great exhaustion, the death occurring with little or no muscular movement, because in such a case the beam could be kept more perfectly at balance and any loss occurring readily noted'. People dying of pneumonia, for instance, were not suitable subjects, as they would 'struggle sufficiently to unbalance the scales'.

The ideal test subjects turned out to be people who were terminally ill with tuberculosis, whose final moments were as inactive as one could possibly imagine. MacDougall found them in the Cullis Free Home for Consumptives in Dorchester, Massachusetts. He wrote in the *Journal of the American Society*

MacDougall's experiments caused quite a stir in the press. (*Washington Post*, 18 March 1907).

SOUL WEIGHT PUZZLES

Hard to Believe Weird Theory and Hard to Discredit.

FACTS MEET NO EXPLANATION

Experiments Show Body Loses Weight at Death and Expounder of New Philosophy Defies Scientists to Solve Riddle—Firm Himself in Belief that Loss Is Due to Spirit Taking Flight.

PLAN TO WEIGH SOULS

Physician Proposes Experiment with Death Chair.

FAIR TEST OF NEW THEORY

Dr. Carrington, of New York, Says Proposed Method Would Do Away with Uncertainty Present in New England Experiments When It Is Said Immortal Part of Man Weighs an Ounce.

A proposed improvement on MacDougall's experiments: using subjects on the death chair (*Washington Post*, 12 March 1907)

The film *21 Grams* (2003) by Alejandro González Iñárritu got its title from the attempts to weigh the soul that were made a hundred years earlier.

for Psychical Research that he had obtained the patients' permission some weeks before their deaths. Even so, there were some people who were sceptical of MacDougall's studies in biological theology. In the case of one of the six research subjects who were weighed, MacDougall complained that the scales had not been adjusted finely enough, and that there was 'a good deal of interference by people opposed to our work'.

MacDougall placed the first dying patient on his scales at 5.30 in the evening. Three hours and 40 minutes later, 'he expired and suddenly coincident with death the beam end dropped with an audible stroke hitting against the lower limiting bar and remaining there with no rebound'. MacDougall had to place two dollar pieces on the scale to rebalance it. These weighed 21 grams.

The next five subjects presented a confusing picture. In two cases, the measurements were unusable, while a third patient's weight dropped after death but remained stable after that. However, in two other cases, the patient's weight decreased before increasing again, while the fifth patient's weight fell, went back up, then dropped again. MacDougall also had difficulty determining the exact time of death.

"by weighing a human being in the act of death"

Yet such details did not shake his belief that he had proved the existence of the human soul. Indeed, he conducted a second experiment that confirmed his finding: 15 dogs weighing 'between 15 and 75 pounds [6.8 and 34 kg]' perished on the scales – all without the slightest loss of weight. In his article in *American Medicine*, MacDougall did not reveal how he managed to persuade the dogs to die on his weighing machine, but in all likelihood he poisoned them. MacDougall professed himself unhappy with this experiment. Not because he found it reprehensible to kill 15 healthy dogs out of scientific curiosity, but because the results could not be compared directly with those of his test subjects. Ideally, the test should have been done on dogs that also were so seriously ill that they were unable to move, but, as MacDougall wrote, 'It was not my fortune to get dogs dying from such sickness.'

Among the scientific community there was a wide divergence of opinion on MacDougall's soul-weighing machine. Some of his colleagues thought the experiments were silly, while others felt MacDougall had made 'the most important addition to science that the world has known' and discussed how his method could be improved. In particular, the use of dying subjects struck them as problematic, since it was perfectly possible that the changes in weight were attributable to the rapid onset of decomposition. 'How much more satisfactory it would be if the subjects were normal men in perfect health,' a New York doctor was quoted as saying in the *Washington Post*. He suggested suspending an electric chair on a set of scales and determining a condemned person's weight before and after electrocution.

MacDougall carried out further experiments, attracting attention once more in 1911 when he claimed to have seen the soul leaving the body in the form of 'a strong ray of pure light'.

The only enduring legacy of this experiment is the weight loss that occurred in the first test subject: for a century, the idea that the soul weighs 21 grams has persisted in popular culture. In 2003, that notion even made it into the movies, when, in a film entitled *21 Grams,* director Alejandro González Iñárritu explored the deeper meaning of life and death.

Source

MacDougall, D. (April 1907), 'Hypothesis Concerning Soul Substance Together with Experimental Evidence of The Existence of Such Substance'. *American Medicine,* pp. 240–243.

1902 Pavlov Only Rings Once

The Russian medic Ivan Petrovich Pavlov is the holder of an unusual record: no set of scientific experiments has had more bands named after them than those that Pavlov conducted at the beginning of the 20th century with dogs. In the 1970s there was a rock band that went by the name of Pavlov's Dog and the Condition Reflex Soul Revue and Concert Choir, while in the 1980s Ivan Pavlov and the Salivation Army made their debut. The 1990s saw not only the bluegrass band called Pavlov's Dawgs, but

madsciencebook.com An educational video about Pavlov.

also a rock group called Conditioned Response, while the new millennium witnessed the appearance of the English folk outfit Pavlov's Cat. Nor were musicians the only ones who lighted on Pavlov when scouting around for a name: 'Pavlov's Dog' is also the name of a communications agency in Ireland, a pub in England, a theatre group in Canada, and a drink in the One World Café in Baltimore, USA, a blend of Kahlúa, Bailey's and milk.

Pavlov was awarded the Nobel Prize in 1904 for his researches into the digestive system. However, this is not the reason why his name is so well known nowadays. Rather, the source of his fame is the basic learning mechanism that he discovered quite by chance in the course of his studies.

While undertaking his research into digestion, Pavlov also became interested in the function of the salivary glands, and in order to observe how they worked in living dogs, he channelled the animals' saliva directly from the gland through a hole in their cheeks into a small measuring

View of one of the experimental booths that Pavlov had erected in his laboratory in order to ensure that the dogs remained completely isolated during his experiments. A system of levers and cables (shown left) enabled the scientist to carry out all the necessary operations for feeding the dogs from outside the booth.

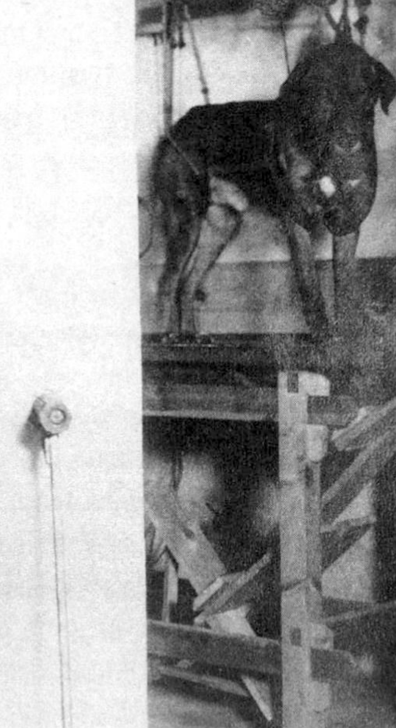

jar. In doing so, he was trying to determine the composition of their saliva when he fed the dogs with different diets. However, a problem soon arose. After the dogs had been fed a couple of times, they began to secrete saliva as soon as they caught sight of the food. At first Pavlov regarded this effect as a disruptive element in his research and so devised ways of putting the food into the dogs' mouths without alerting them in advance. Yet it transpired that the animals associated even quite subtle signals with feeding. The mere sight of the researcher or the sound of his footsteps were enough to stimulate the flow of saliva.

Ivan Petrovitch Pavlov was awarded the 1904 Nobel Prize for Medicine. However, his lasting fame rests on the later experiments on conditioning that he conducted using dogs.

It was not long, however, before Pavlov stopped regarding this phenomenon as a flaw in his experimental method and came to look on it as a new area of research in its own right. He conducted experiments in which he controlled the signals that occurred prior to feeding. Five seconds before feeding, a metronome was set in motion or an electric bell rung. After a few such couplings – in the case of the bell, a single ring was enough – the dogs' saliva began to flow as soon as they heard the signal. The dogs had learned that they would be fed after the bell was rung.

'And then, instead of feeding me, he rang a bell.'

Because the dogs interpreted even the faintest cues from their surroundings as signals for impending feeding, Pavlov had a new building with soundproofed rooms constructed in St Petersburg, in which he could perform all the necessary operations remotely via a system of cables and levers.

The fundamental learning mechanism that Pavlov discovered through his experiments is known as 'classical conditioning'. This involves pairing a natural combination of stimulus and response (i.e. food and

No scientific experiment has ever had so many bands named after it. Below, the album *Pavlov's Dog* (1997) by the American rock band Conditioned Response.

salivation) with a new stimulus (i.e. a bell). In doing so, the new stimulus can only ever trigger an innate behaviour, though admittedly this can occur in almost any combination one could think of. On the other hand, the question of how new behaviours are learned was only investigated 30 years after Pavlov's experiments by the American psychologist B.F. Skinner, using the so-called Skinner box (see p. 97).

During his experiments, Pavlov also discovered how conditioning can be unlearned. It was only necessary to ring the bell a couple of times without then feeding the dog for the animal to forget the connection. This was the basic principle behind the later development of behavioural therapy, which involved patients being confronted under controlled conditions with precisely those situations that usually triggered anxiety, for example. This technique was designed to break the association between certain situations and anxiety.

Today, everyone is familiar with the concept of Pavlov's dogs. Cultural commentators have used them to symbolize the common herd in Western industrial societies, who have allowed themselves to be trained by advertising into 'consumer animals' and who display predictable purchasing reactions when confronted with particular stimuli.

Unlike Pavlov himself, who has gone down in history as one of the most famous scientists of all time, the bands named after him never made the breakthrough – or at least haven't done so yet. The one that came closest was the rock group Pavlov's Dog and the Condition Reflex Soul Revue and Concert Choir, who were rechristened simply Pavlov's Dog in 1973 and paid $600,000 for their debut album, which at the time was the largest ever advance for a record in the USA. But three years later their record company dropped them, and the musicians found themselves broke and in dispute with one another.

Source

Pavlov, I.P. (1927), *Conditioned Reflexes: An Investigation of the Physiological Activity of the Cerebral Cortex.* Oxford University Press.

madsciencebook.com **An animation about the theory of classical conditioning.**

1904 The Horse Whisperer

In the summer of 1904 a cobbled courtyard in the north of Berlin played host to a remarkable spectacle. There, surrounded by tenement blocks, the retired teacher Wilhelm von Osten would give demonstrations – always at midday – of his horse Hans's extraordinary skills. Hans could add fractions, count people, recognize pictures and tell the time; what's more, he also had perfect pitch and had memorized the calendar for the entire year. Newspapers worldwide carried reports about the horse prodigy. *The Mexican Herald* even speculated that it surely wouldn't be long before Hans did his first tour of America. Yet despite being feted in popular songs and having toys and alcoholic drinks marketed under his name, Hans is not remembered nowadays for his supposed cleverness but rather for the experiments that disproved his intelligence.

Curious onlookers gather to see one of the midday performances given by Clever Hans in the back courtyard of number 10 Griebenowstrasse in Berlin.

Wilhelm von Osten had spent four years teaching Hans like a schoolboy. The horse stood in the courtyard in front of a blackboard, and von Osten taught it counting with the aid of an abacus, used a board with letters on it to teach it reading and a toy harmonica to teach it music. As the horse couldn't speak, it answered either by nodding or shaking its head or stamping its hoof on the ground. Letters, notes on the musical scale and even the names of playing cards were converted into numbers and registered with hoof beats. Thus, an ace was one hoof beat, a king two, a queen three, and so on. This teaching method was 'well thought out and could probably be put to practical use in educating Hottentots' as von Osten later attested.

The scientific world began to sit up and take notice of Hans. Respected figures such as the circus impresario Paul Busch, the zoo director Ludwig Heck, the veterinarian Dr Mietzner, and Carl Stumpf of Berlin University, one of the foremost psychologists of the age, came to see the horse perform. They were sufficiently impressed by Hans's abilities that they signed a curious affidavit on 12 September 1904. The Hans Commission, as the 13 signatories of this document styled themselves, certified that von Osten had not employed any tricks. In their opinion, Hans had received no signals, either conscious or subconscious, from his trainer. They concluded that 'this is a case that differs fundamentally from all previous, ostensibly similar instances'.

Lessons in counting for the world's most famous horse, Clever Hans. A sophisticated experiment showed what really lay behind the animal's remarkable skills.

Indeed, as one enthusiastic spectator put it, it seemed as though the only thing that distinguished the stallion from a sentient human being was the capacity for speech. A seasoned educationalist rated Hans as being at the developmental level of a 13- or 14-year-old child. The few mistakes that Hans made were interpreted as 'signs of wilfulness and independence that one might almost take to be a sense of humour'. Some zoologists went so far as to regard Hans as living proof of the essential similarity between the human and animal spirit.

This phenomenon, as the affidavit concluded, 'warranted serious, in-depth scientific investigation'. The psychologist Stumpf entrusted this task to his student Oskar Pfungst. Although he found out that Hans's supposed intelligence didn't actually amount to much at all, what he discovered in the process was no less astonishing than a horse that could count.

Teacher and pupil: Wilhelm von Osten and Hans with the abacus, the spelling board and other teaching aids.

Pfungst's first experiment was designed to ascertain whether Hans really was solving the tasks he was set without human help. If this was indeed the case, then it would have made no difference whether the person conducting the experiment knew the answers to the questions or not. Pfungst asked Hans to respond with the requisite number of hoof beats to a series of numbers he showed the horse on cardboard placards. Alternately, he either showed the card only to the horse or looked at it himself as well. The result was unequivocal: when Pfungst knew the numbers, Hans's strike rate was 98 per cent, but when he didn't, the horse only got 8 per cent correct.

To test Hans's ability to count, two people would each whisper a number into the horse's ear for it to add up. These conditions meant that only Hans could know the result. He failed abjectly. Pfungst

'No horse can be expected to put up with all this indefinitely,' said Clever Hans, casting a sideways glance at the Parliament Building, 'so, farewell Berlin.' This cartoon appeared in a 1909 issue of the satirical magazine *Kladderadatsch*, marking Hans's departure from the German capital.

was now certain that the stallion 'must be picking up certain cues from his environment'. This finding was in itself astonishing, since unlike counting animals in circuses, whose trainers gave them signals, von Osten wasn't even present at many of the experiments. The only person who could have conceivably been giving Hans some sort of signal was the chief investigator – Pfungst himself! But he wasn't remotely aware he was doing so.

Was the questioner really giving hints unwittingly? Pfungst put blinkers on the horse to restrict its vision. These experiments were difficult because the horse still persisted in trying to catch sight of Pfungst, even breaking loose in these frantic attempts. Nevertheless, the conclusions were undeniable: when Pfungst wasn't in Hans's field of vision, the horse was no longer able to give the correct answers. Clearly, Hans could somehow divine the answer from the demeanour of the questioner. But how? After all, Pfungst had done his utmost to stay neutral.

Hans's supposed abilities were also widely reported in American newspapers. This headline appeared in the *Stevens Point Daily Journal* of 13 October 1904.

A Marvel in Animal Education in Germany

"The Wonder Horse" of Berlin Described by Prof. Amos W. Patten of Northwestern University.

After making minute observations, Pfungst came to the conclusion that Hans must be taking his cue from the tiniest involuntary movements that he, Pfungst, made with his head. Anyone asking the horse a question nodded ever so slightly when glancing down at the horse's hoof. For Hans, this was the signal to start stamping his hoof on the ground. As Hans got near the required number, the questioner would look up again, and the horse promptly stopped.

Pfungst devised further experiments to test this hypothesis.

For instance, he tried setting Hans tasks from various distances. The further away he was, the more unreliable the answers became. Hans couldn't read the questioner's body language as accurately from a distance. Pfungst also asked Hans questions to which the answer was 1. If his assumption was correct, the horse would find these calculations the hardest to do, since the experimenter would be signalling the animal to start and stop stamping at virtually the same time. And so it turned out – of all the numbers, the one that Hans had the most trouble with was 1.

But the real clincher came when Pfungst showed that Hans began to stamp his hoof even when the experimenter leaned forward slightly, but without having asked a question.

Yet Pfungst wasn't content simply to leave it at that. In November 1904 he asked 25 people to visit him at the Institute of Psychology in Berlin, one after the other. The subjects had no idea what the tests were about, as Pfungst asked them to think of a number, which he would then guess by rapping on the table top an appropriate number of times. In his famous book *Der kluge Hans* ('Clever Hans'), Pfungst later described this test in the following terms: 'Instead of a dumb horse, I was now working, as it were, with talking horses.' In 23 out of 25 cases, he managed to read his subjects' involuntary body language and guess the number correctly.

Pfungst's studies underlined one of the most significant disruptive factors that can affect the outcome of any experiment: the investigator's expectations. As many later studies were to demonstrate, researchers unwittingly skew the results of their experiments in favour of their own presuppositions. Unconsciously, Pfungst was signalling to Hans when he wanted him to stop stamping. In modern psychology, this phenomenon is referred to as the 'experimenter effect' and must always be taken into account when designing an experiment. Pfungst's work is so renowned that even today symposia are held to discuss the 'Clever Hans Phenomenon'.

But was Pfungst the real brains behind the experiments? The German psychologist Horst Gundlach has raised doubts on this score, having identified certain discrepancies in the traditional version of events. For example, for years Pfungst had been working on a doctoral thesis without ever managing to finish it. Why couldn't he simply have submitted his work on 'Clever Hans' as his thesis? And how did Oskar Pfungst, who wasn't even a research assistant, come to be the sole author of the book? It is normal practice for a professor to add his name to his students' work. Pfungst never wrote another book thereafter and hardly even published any papers. Gundlach thinks that large parts of the book weren't in fact written by Pfungst but by his boss Carl Stumpf. However, Stumpf no longer wished to be associated with the horse, since when Hans was exposed both his colleagues and the press mocked him for having believed in the horse prodigy's abilities for several months.

The experiments cost Wilhelm von Osten his livelihood. As the results began to emerge, he wrote to Carl Stumpf requesting that researchers kindly stay away in future. There is no question that von Osten wasn't a swindler, but was merely, like Pfungst, conveying unconscious signals through his body language. Among the questions that he asked Hans during the performances in Berlin were ones relating to whether the horse liked or disliked certain people. He once pointedly asked: 'Do you like privy Councillor Stumpf?' Hans shook his head.

Wilhelm von Osten died on 29 June 1909. On his deathbed he cursed his horse, whom he held responsible for the unfortunate turn his life had taken, saying that he hoped it would 'end its days hauling a cement wagon'.

He bequeathed Hans to Karl Krall, a businessman from the town of Elberfeld. Krall set up a 'Stable for the Purposes of Instruction', where he also taught the horses Muhamed and Zarif. All the horses in Krall's stable were pressed into military service at the outbreak of the First World War.

Source

Pfungst, O. (1907), *Der kluge Hans. Das Pferd des Herrn von Osten* ('Clever Hans, Mr Van Osten's Horse'). Johann Ambrosius Barth.

1912 Happy Birthday, Dear Cells!

17 January was a red-letter day in Alexis Carrel's laboratory at the Rockefeller Institute in New York. On this day every year, the lab's employees would gather in front of a sealed Pyrex flask and sing 'Happy Birthday'. The recipients of their felicitations were cells from a chicken's heart, which Carrel had put into a culture medium on 17 January 1912. In no time, these 'immortal cells', as they were known, were so famous that the *New York World Telegram* newspaper would enquire after their wellbeing every January.

Carrel wasn't the first person to try to keep cells alive outside the body. The fact that it was his name, first and foremost, that later became associated with such attempts was down to his technical flair and his predilection for showmanship. What impelled him to undertake the experiment with the chicken's heart was criticism by fellow scientists, who had cast doubt on Carrel's successes in cultivating thyroid-gland and kidney cells. Of the numerous cell cultures that he began cultivating under various different growth conditions, it was the one numbered 725 that provided him with a major breakthrough at the start of 1912. He placed small pieces from the heart of a chicken embryo into a solution of blood plasma and distilled water at a temperature of 39°C; the cells began to divide.

This was the container in which Alexis Carrel grew his renowned 'immortal' cells. They are reputed to have survived for 34 years.

After a few days, the heart tissue was cut up into tiny pieces and cleaned. Then, suspended in a variety of subcultures, the pieces were placed into a new culture medium. Cell cultures such as this were supposed to help answer one of the burning questions of biology at that time, namely, why do we age? Is it the case that individual cells in the body somehow become infirm, or is it rather that our entire system of cells is responsible for our gradual decline?

Carrel regarded his cells as immortal. 'Aging and death are not inevitable phenomena,' he wrote a year before he even began his experiments. The object of his work was to find out the conditions 'under which the active life of a tissue outside of the organism could be prolonged indefinitely'. As soon as he had discovered what these were, he averred, 'we shall be able to create living beings'.

A lab assistant injecting fresh culture medium into the cell cultures. It later turned out that the 'immortal' cells were being given occasional help to survive.

The lasting fame of Carrel's immortal calls was assured when, shortly after embarking on his experiments, he was awarded the Nobel Prize for his work on vascular surgery. The *Rural Weekly*, published in St Paul, Minnesota, made this its title story on 24 October 1912: 'He keeps hearts alive in test tubes and wins $39,000 Nobel Prize.' The article went on to make the following claim: 'it is said that he [Carrel] could, by taking various parts of different animals, build up a new living creature'. Of course, this was hugely exaggerated, and similar outrageous claims were made about the 'immortal' cells. On the way from the laboratory to the newspaper columns, the millimetre-long pieces of tissue had miraculously metamorphosed into a beating chicken heart in a preserving jar mounted on a white marble plinth. Allegedly, the heart had to be trimmed back from time to time to prevent it from overrunning the laboratory.

In 1940, when Carrel left the Rockefeller Institute, the *New York World Telegram* published an obituary for the cells. However, they weren't dead – a colleague of Carrel's continued to keep them alive. They were only finally abandoned to their fate on 26 April 1946. When biologists subsequently discovered that cells die off outside the body after dividing a certain number of times, they were faced with a puzzle: by rights, Carrel's cells shouldn't have survived for 34 years!

All attempts to replicate Carrel's experiment ended in failure. Although cells could be kept alive outside the body, none lived as long as those in Carrel's Pyrex flask. According to the biologist Jan A. Witkowski, who studied the case in detail, there were three possible explanations: first, Carrel's cells had mutated to such an extent that they had acquired the ability to divide indefinitely like cancer cells; second, that new cells had inadvertently been added to the old cells every time the culture medium was refreshed; or third, that the old cells had long since passed

UNHATCHED CHICK'S HEART IS BEATING AFTER TEN YEARS

NEW YORK Jan. 17.—Part of the heart of a chicken that never was hatched was beating today, the tenth anniversary of its removal from the embryo and isolation by Dr. Alexis Carrel of the Rockefeller Institute.

The tissue fragment is still growing and its pulsations are visible under the microscope, Dr. Carrel said. It grows so fast that it is sub-divided every forty-eight hours.

Finally, on their tenth birthday, the cells made it onto the front page; this report appeared in the Reno Evening Gazette on 17 January 1922.

Alexis Carrel was awarded the Nobel Prize for Medicine in 1912. His daring organ-transplant procedures earned him the reputation of being something of a magician among surgeons, as in this cartoon of 1914.

away and Carrel or his assistants had constantly added new cell cultures and passed them off to the outside world as the old ones. If this latter explanation were true, then one of the most celebrated experiments in the history of biology was nothing but a fraud.

As early as the 1930s there were indications that this might indeed be the case. Rumours began to circulate that there was something not quite kosher about Carrel's cells. One of his assistants is even supposed to have told a visitor to the lab: 'Well, Dr Carrel would be upset if we lost the strain, so we just add a few embryo cells now and then.'

As a later researcher wryly remarked, 'The ultimate effects of the aging process have made it impossible for Carrel to respond in his own defence'. Carrel died on 5 November 1944 in Paris – two years before his 'immortal' cells.

Source

Carrel, A. (1912), 'On the Permanent Life of Tissue Outside the Organism'. *The Journal of Experimental Medicine* 15, pp. 516–528.

1914 The Building of the Tower of Banana

It's an extremely iconic image: a chimpanzee balances on three crates stacked on top of one another and reaches up for a banana. For generations of psychology students, it symbolized the tale of the intelligent ape which, in a sudden flash of inspiration, set about piling one crate on top of another until it could reach the fruit that would otherwise have been beyond its grasp. But in fact the truth was somewhat more complex.

The German psychologist Wolfgang Köhler, who devised this mental puzzle for apes, landed on Tenerife at the end of 1913 to take over the running of the anthropoid station on the island. He planned to stay only a year, but the First World War intervened, and in the event that one year turned into six.

During this period Köhler conducted a series of elegantly designed experiments on the intelligence of the anthropoid apes. In doing so, he became convinced that chimpanzees exhibited 'insightful behaviour like that encountered in human beings'. This finding set the minds of contemporary evolutionary biologists at ease. It had not been long since Darwin had posited the theory of natural selection, and the generation of biologists that came

Proof of intelligence? The female chimpanzee called Grande heaped up wooden crates in order to get to a hanging banana.

after him sought evidence throughout the physical world to support this theory. The similarity between the human body and that of apes was proof of their relatedness, yet Darwin was convinced that humans and apes must also share mental characteristics. Exactly how close such affinities were was what Köhler set out to discover.

On 24 January 1914 he led six of his chimpanzees into a room 6 ft (2 m) high, hung up a banana in the corner and placed a wooden crate in the centre of the room. Then he waited. All the chimps tried in vain to get at the banana by jumping. 'However, Sultan soon gives up,' wrote Köhler, 'then paces uneasily round the room, before stopping suddenly in front of the crate, grabbing hold of it and quickly shoving it directly towards the target. When he is still half a metre short (in a horizontal direction), he clambers on top of it and, leaping up with all his might, snatches down the desired object.' Sultan had successfully solved the problem. He had acted in an impulsive and decisive manner, as though he had had a sudden brainwave.

Köhler was even more surprised to find that his chimpanzees took ages to solve the next problem he set them. This time the banana was hung even higher, so that the chimps could only get to it if they piled one crate on top of another. Köhler observed that, for the ape, what it was being asked to do fell into 'two clearly defined parts, one of which it was able to solve really easily, while the other caused it untold difficulties'. The simple task consisted of pushing a crate until it was underneath the banana, whereas the difficult one came in placing a second crate on top of the first. Köhler was at loss to explain this 'remarkable fact', since things were quite different where humans were concerned. As soon as a person realizes that he can reach the banana by pushing a crate underneath and climbing on top of it, then he realizes he can accomplish the same feat with two or three crates piled on top of one another when the banana has been hung higher. For him,

Source

Köhler, W. (1921), *Intelligenzprüfungen an Menschenaffen* ('Intelligence tests on anthropoids'). Springer.

'the placing of one building block on top of the other is merely a repetition of the first action of placing one crate on the ground'. This was not the case for the chimp.

Grande, the chimpanzee in the picture, struggled repeatedly with the second crate. Although she eventually managed to construct a small tower, she kept making the same mistake – for years on end. And even after several successful attempts, she suddenly found herself completely at a loss once more to know what to do with the second crate. Köhler concluded that chimpanzees had no understanding of how the crate towers they built balanced and that 'they solved almost every "question of balance" that arose during construction not intuitively but purely through a process of trial and error ...'

Köhler's experiments are nowadays regarded as classics, and are still being conducted in modified form. Even so, it is hard to determine what they tell us about the affinity between human beings and chimpanzees. After all, similar behaviour on the part of humans and chimps need not necessarily derive from similar thought processes.

1917 Dr Watson's Divorce

Mary Ickes Watson is surely one of the few women ever to have obtained a divorce because of a scientific experiment. Newspapers at the time were full of reports about the messy separation. John B. Watson, Mary's husband, was a prominent psychologist. In 1915 he was elected president of the American Psychological Association, and four years later was voted the most handsome professor at Johns Hopkins University in Baltimore by his students. Therein, perhaps, lay one of the reasons for the problems he was to encounter: he was simply too good-looking.

During the First World War, Watson had seen information films that were shown to American servicemen before they travelled to vice-ridden Europe. Drastic images of sexually transmitted diseases warned the men against having contact with prostitutes. At the end of the war, Watson showed these same films to civilians and doctors and interviewed them afterwards. He concluded that many doctors regarded sexuality as immoral per se and consequently saw it as a kind of disease.

Source

Magoun, H.W. (1981), 'John B. Watson and the Study of Human Sexual Behaviour'. *Journal of Sex Research* 17, pp. 368–378.

For Watson, this was a sure sign that research into sexuality should no longer be left to the medical profession alone. It was high time that psychology began to investigate human sexual behaviour. After Watson's death, his friend and colleague Deke Coleman revealed that Watson led by example in this matter, carrying out experiments of his own. But since Coleman has also died in the interim and can no longer be quizzed about this, several of Watson's biographers remain sceptical as to whether these experiments really took place.

According to Coleman, from roughly 1917 onwards, Watson, then aged 39, conducted experiments of the most intimate kind with the student Rosalie Rayner, 20 years his junior. In these experiments he recorded his and her physical reactions as they had sexual intercourse together. James V. McConnell, who was the first to reveal this in his 1974 book *Understanding Human Behavior*, reckons that Watson's observations on the sexual act were the earliest of their kind ever to be made. Admittedly, Watson's wife wasn't exactly well placed to appreciate her husband's scientific achievement. Growing suspicious, she began to look for proof. Rayner came from a prominent family and still lived with her parents. When they invited the Watsons to come to stay with them, in the course of the evening Mary Ickes Watson feigned a headache and asked to lie down for a moment. But instead of resting in a first-floor room, she rifled through Rosalie Rayner's room and came upon love letters written by her husband, which promptly found their way into the *Baltimore Sun*. In them, he wrote '… every cell I have is yours, individually and collectively … I can't be more yours than I am, even if a surgical operation made us one'.

At the insistence of the president of Johns Hopkins University, Watson was forced to renounce his professorship after the divorce, and worked thereafter in advertising. He married Rayner and lived with her until her premature death in 1935. Before leaving the university, he conducted one of the most famous experiments in psychology: the conditioning of 'Little Albert' (see below).

Following Watson's sex experiments – the findings of which his ex-wife had destroyed (along with all the legal papers relating to the divorce) – it was another Ten years before another couple was observed for scientific purposes during the sex act. The results were more satisfactory, in all possible respects (see p. 92).

1920 Putting the Wind up Little Albert

Was his name really Albert? Albert B., as he was called in the study? If so, then it might be possible to track him down now. He'd be in his late eighties. But in all probability he wouldn't even be aware that he is the famous 'Little Albert' whose scream is familiar to every student of psychology. He was just nine months old when he played the lead role in a film in which his co-star was a white rat. Perhaps, whoever he is, he still suffers from a pronounced fear of white rats.

The most famous child's scream in psychology: John Watson (right) and Rosalie Rayner experiment on the generalization of fear with 'Little Albert'.

Albert's mother was a wet nurse in the Harriet Lane Home for Invalid Children, where Albert spent much of his time. The psychologist John B. Watson and his assistant Rosalie Rayner (see above) had good reason for choosing this baby in particular for their experiments. 'He was on the whole stolid and unemotional,' the two researchers later wrote. This emotional equanimity persuaded them to conduct the tests with Albert: 'We felt that we could do him relatively little harm by carrying out such experiments as those outlined below.'

John Watson wanted to apply the findings that Pavlov had obtained with his dogs (see p. 63) to humans. Watson later became the founder of behaviourism, a movement within psychology that confined itself to the study of human behaviour, while refusing to speculate on internal, and therefore objectively inaccessible, processes within the mind. Watson was convinced that human behaviour could be understood only as a sequence of reactions to external stimuli.

However, there was one snag with this theory. Only a very limited range of innate reactions had been observed in babies – for example, fear of loud noises or anger at having their movements constricted. By contrast, adults exhibited such innate reactions to every conceivable type of stimulus, be it other people, things, or events. Watson and Rayner concluded from this that 'there must be some simple method by means of which the stimuli which can call out these emotions and their compounds is greatly increased'. And this method, he reckoned, was conditioning.

Original footage of little Albert.

The first test with Albert took place when he was 8 months and 26 days old. Watson hammered on a steel rod hanging behind the child's back. Albert reacted immediately: 'The child started violently, his breathing was checked and the arms were raised in a characteristic manner. On the second stimulation, the same thing occurred, and in addition the lips began to pucker and tremble. On the third stimulation the child broke into a sudden crying fit.' This, then, was the innate connection between noise and anxiety that Watson aimed to exploit to teach the child to fear new things.

Only a single paragraph in Watson and Rayner's report of the experiments betrays that they were troubled by any pangs of conscience regarding their actions. They assuaged their concerns, however, with the thought 'that such attachments would arise anyway as soon as the child left the sheltered environment of the nursery for the rough and tumble of the home'.

When Albert was 11 months and 4 days old, Watson taught him to fear a white rat. Taking the animal from a basket, he placed it down in front of the sitting child and let it run free. Albert showed absolutely no fear and stretched out his hand towards the rat. But the instant he touched it, Watson started hammering on the steel bar: 'The infant jumped violently and fell forward, burying his face in the mattress. He did not cry, however.' When the baby tried a second time to touch the rat, Watson began hammering once more. The child started to cry. A week later, Watson and Rayner resumed the experiment. Whenever Albert touched the rat, they made a din with the steel rod. They repeated this twice, three times, four times. In between, they simply showed Albert the rat to see whether they had achieved their end yet. After the seventh occasion on which they combined the rat and the noise, Albert began to bawl just at the sight of the rat. Watson and Rayner had succeeded in creating in the child an association between fear of loud noises and a new stimulus – the rat.

Five days later, Watson attempted to find out whether Albert would transfer his fear of the rat to other animals and objects. And indeed, the child now exhibited fear at the sight of a rabbit, a dog, a sealskin coat and – somewhat less markedly – at cotton wool, hair and a Santa Claus mask. As control objects, Albert was repeatedly given building blocks, which elicited no fearful reaction from him and which he began to play with straight away.

The infamous film that Watson made of Little Albert made the experiment so popular that it has become part of the folklore of psychology, and is recounted in a number of inaccurate versions. For instance, several textbooks claim that Watson showed Albert a cat, a muff, a white fur glove and a teddy bear. Albert's reactions have also been liberally reinterpreted to fit particular theories. Moreover, certain authors describe in detail how Watson rid Albert of all the fears he had conditioned in him before the end of the experiment. Yet, in fact, this is precisely what he didn't do. This is particularly astonishing, since he knew in advance exactly when Albert and his mother were going to leave the children's home and also because he was well aware of the potential consequences of his experiments. When Watson published the results, he wrote that 'these responses in the home environment are likely to persist indefinitely, unless an accidental method for removing them is hit upon'.

Not long thereafter, Watson was dismissed by the university for another experiment, one in which he had got rather too intimate with Rosalie Rayner (see p. 77). He subsequently wrote a popular book on child education, in which he advised parents not to give their children too much love or attention. Forty years after the experiment with Little Albert, the psychologist Harry Harlow demonstrated through a series of cruel experiments with monkeys how wrong Watson was to suggest this (see p. 154).

In the event that Little Albert is still alive, he can console himself with the fact that his fame has even spread into the realm of popular music. In 2002 the Texan rock band Crevice released an album entitled

Source

Watson, J.B., and Rayner, R. (1920), 'Conditioned Emotional Reactions'. *Journal of Experimental Psychology* 3 (1), pp. 1–14.

Lullaby for Little Albert. There is a picture of Albert on the back of the CD booklet, and the band showed excerpts from the film of the experiments during their live shows. But it is debatable whether Albert would actually have been gently rocked off to sleep by this 'lullaby' – Crevice are known for their tuneless experimental techno.

1923 Male Urges in Female Bodies

Walter Finkler chose his experimental animals with care. The jet-black water beetle Hydrophilus piceus was not only easy to keep but also had a sex life that was ideally suited to the purposes of research, in that H. piceus did not copulate 'at night or in any other clandestine manner', as Finkler put it. This was a key consideration, since Finkler, a researcher from the Experimental Biological Station in Vienna, intended to use the beetle's behaviour during copulation as a 'benchmark for sexual instincts'.

For some time, Finkler had been undertaking transplantation experiments with beetles. The procedure was straightforward and – as critics of his achievements remarked – 'scarcely to be surpassed in its sheer brutishness'. Finkler would starve one of the insects for two or three days before anaesthetizing it with sulphur ether, cutting its head off with scissors, and then transplanting the head onto the body of another insect that had had its head removed. The transplantee was held in place until the head was assumed to have grown on.

Finkler claimed that he had even used this method to successfully swap the heads of different species of beetle. As he wrote in one of his papers on the subject, 'The water beetle swims about with the head of a great diving beetle [Dytiscus marginalis] as naturally as if it had never had any other head throughout its entire life.'

It was only a matter of time before the question arose as to what happened when the heads of male and female water beetles were switched. Would the head or the body determine their sexual behaviour?

However, before Finkler could solve this puzzle, he had to answer another question, namely, 'Might homosexuality not also exist among water-beetle populations?' For example, in the event of an 'abnormal exercise of the sexual urge' doubts would arise as to whether it was a heterosexual female head or a homosexual male body that was governing the behaviour of the composite insect.

Although such 'perversions' did occur among a variety of beetles, Finkler had never noticed it in two years of observing Hydrophilus. The experiment could therefore begin. The biologist operated on the insects, placed various permutations of them in tanks, and witnessed a phenomenon that he later skilfully reworked into a soundbite: 'Male urges in female bodies'. As he went on to explain, 'Females with male heads prepared for copula, acting in the process as though they were males.' In describing the behaviour of the females, Finkler thought it fitting to introduce his own conception of womanhood: 'And what of the female – the true female, that is – that had been seduced into fornication in this way? Not only did she resign herself to everything that was going on but even took pleasure in it, adopting the attitude: "Do with me what you will!".' Yet in describing sexual intercourse, he did concede that even he, an experienced scientist, wasn't party to the innermost experience of a female beetle. He described the typical motions of rejection made by the female thus: 'Who can say whether they are genuine or skilfully dissimulated? How can we hope to ascertain this in female beetles, when we can't even tell what's what with the female of the human species!'

Walter Finkler's bizarre beetles. Top right, a normal Hydrophilus sp. and bottom left, a normal Dytiscus sp. All the rest have transplanted heads.

Many researchers tried in vain to repeat Finkler's experiments. In 1924 the entomologists Hans Blunck and Walter Speyer arrived at the following conclusion after 52 pages of painstaking description of their own experiments: 'With regard to the claims of our Viennese colleague, which run counter to all experience, there is no further reason for scientific investigation to concern itself with him or his publications.'

Finkler's writings do indeed contain a number of clear inconsistencies. It is possible that he fell victim

Source

Finkler, W. (1923), 'Kopftransplantation an Insekten' ('Head transplants on insects'). *Archiv für mikroskopische Anatomie und Entwicklungsmechanik* 99, pp. 104–133.

to the 'experimenter effect' (see p. 67), but it is more likely that he was a fraud.

1926 Boxes Help People Think Outside the Box

This is one of very few experiments that have made the leap out of the psychology lecture theatre and into books of puzzles. You can try it for yourself. Place three cardboard boxes (about the size of a matchbox) on the table, each containing drawing pins, one small candle and matches. Your task now is to attach the candles to a door at eye level. Well then? – the answer's really quite simple: you attach the boxes to the door with the drawing pins and then use them as candle supports.

The key step in the process is to rethink the function of the boxes from being 'containers' to 'supports'. This is what many researchers nowadays regard as the secret of creativity – the ability to accord something a purpose for which it wasn't originally intended; or, as psychologists put it, to free oneself from 'functional fixedness'.

The candle exercise was just one of many that the German psychologist Karl Duncker set his test subjects. The method was a little more complicated than in the brief description above. As well as the three boxes, there were also other objects on the table, which had no bearing on the solution of the problem. In the instructions he gave his subjects prior to the experiment, Duncker explicitly indicated that they were free to use any of the objects in front of them and that they should think aloud while solving the problem. In this way, he wanted to observe their train of thought.

He devised two versions of the experiment: in one, the boxes contained drawing pins, candles and matches, while in the other the boxes were empty and the drawing pins, candles and matches were lying on the table. Out of seven test subjects, all solved the problem when the boxes were empty, but only three succeeded when they were full. As Duncker had suspected, the test subjects found it easier to disengage the empty boxes from their original function as containers than the full ones, which were, after all, still performing that function.

Source

Duncker, K. (1935), *Zur Psychologie des produktiven Denkens* ('The Psychology of Productive Thought'). Springer.

Duncker then tried to find out more about what conditions might make it easier for a person to 're-centre' – that is, to mentally disengage a thing from its original function. If the contents of the boxes had nothing to do with the solution to the problem – Duncker filled them with buttons, for instance – this made re-centring easier. Specific instructions also helped, such as 'In solving the problem, use the drawing pins and another object that can easily be affixed to the door with the pins.'

Duncker published the results of his experiments in 1935 in his book *The Psychology of Productive Thought*, which nowadays is regarded as one of the key texts of cognitive psychology. He was 32 years old at the time. But because he had strong sympathies with the Communist Party, his appointment as a lecturer was blocked twice. Five years later, afflicted by severe depression, he took his own life.

1927 Assembly by Moonlight

The experiments with light that started it all should really only have proved what any rational person long since knew to be the case: people work better in better light. And yet the highly equivocal results that emerged provided the springboard for the best-known experiments in occupational psychology. Their interpretation is still the subject of controversy even today.

In the 1920s, manufacturers of electrical appliances and light bulbs claimed that electric light prevented accidents, protected people's eyesight and increased productivity. They planned to demonstrate the advantages of electric lighting to their customers by carrying out a series of methodical tests.

One such experiment took place in 1924 at the Western Electric company's Hawthorne Works in Chicago. The procedure was simple: the lighting in various compartments was systematically altered and productivity measured. The conclusion that better lighting led to improved output turned out to be false. Although, over the course of the experiment, the three test groups steadily increased their rate of production, this effect was independent of the strength of the light. What's more, the control group, which didn't even have the benefit of electric lighting, also upped its output.

A partial experiment produced a particularly odd result: the

The T-room at the Hawthorne Works: this was supposed to be where the mystery of productivity was finally solved. Instead, the results of the experiment are still being argued about today.

researchers had two employees work in a cloakroom in extremely poor lighting conditions. Despite this, they managed to keep up their rate of productivity, or even increased it. Their output only started to tail off when the luminous intensity reached 0.06 candelas, the equivalent of a night with a full moon.

This episode is still presented to students of occupational psychology today. The usual line is that researchers then were completely at a loss and thrashed around for ages trying to find an explanation until they finally hit on the idea of taking psychological factors into account. In other words, they are portrayed as naïve mechanics who were trying to study the piece-workers in the factory as isolated machines.

In actual fact, the scientists conducting the studies knew right from the outset that other factors besides the change in lighting would have an impact on the data. And where the test in the dingy cloakroom was concerned, they had even correctly predicted that well-meaning workers would make every effort to maintain their output even under the most unfavourable conditions.

When the science historian Richard Gillespie came to review the original data from all the experiments in researching his book *Manufacturing Knowledge* at the beginning of the 1990s, he unearthed further inconsistencies. In an attempt to present their study as a perfect sequence of hypotheses, experiments and results, the authors of the official publication had altered the chronology of events, suppressed contrary interpretations and retrospectively presented insights that had been arrived at only gradually as sudden flashes of inspiration.

According to Gillespie, the 1939 book *Management and the Worker*, which gives a 600-plus-page account of the lighting experiments, 'has misled generations of social scientists'. The prevailing interpretation of the experiments in textbooks did not proceed directly from the data, but rather had been provided by the book's authors.

Yet in addition, all the numerous new interpretations that had appeared in the 70 years since the conclusion of the experiments had only served to confuse matters rather than clarify them. As one expert said resignedly, 'We shall never know exactly what happened at Hawthorne.'

No concluding report was ever compiled on the lighting experiments. The electrical industry evidently had no interest in publishing results from which they could derive no benefit. Instead, the decision was taken in the Hawthorne Works to begin a series of tests that have gone down in the history of the social sciences as the 'Hawthorne experiments'. They were designed to answer a broader spectrum of questions, such as: What attitudes do workers have to their jobs? Why does productivity go down in the afternoon? And do breaks have a positive effect on the workforce?

For these studies, a test room was set up in the factory, with six workstations for the assembly of the R-1498 relay. The R-1498 was an electromagnetic switch for use in telephone exchanges. It consisted of 32 components, which a female worker could assemble in around a minute. The finished relays were dropped into a sloping chute, where they slid down into a box to be counted.

The social scientist Elton Mayo was called in when problems arose with the interpretation of the data.

The six women in the test room were an entirely separate unit. Their rate of pay was no longer linked to the performance of an entire department with several hundred employees, but was calculated according to the productivity of just their small group. The researchers were afraid that, without this financial incentive, they might not give their full cooperation, which could have jeopardized the experiment.

Yet precisely by putting this measure in place, the scientists managed to jeopardize their own experiment. It was later estimated that the pay scale they introduced for organizational reasons unintentionally exerted a major influence on the results.

Opposite the work bench where the women were assembling the relays sat Homer Hibarger, somewhere between a supervisor and an experimenter, who monitored their work and their hours and noted down a host of details on the experiment.

But this still didn't satisfy the scientists' researching zeal. The six young women, all aged between 15 and 28, had to undergo a medical examination every month. Hibarger used these examinations to find out details about their private lives, including the timing of their menstrual cycles.

The first breaks were introduced in 1927. Initially, these consisted of two 5-minutes breaks in the morning and the afternoon, which then became two 10-minute breaks and six 5-minute breaks, and finally ended up as 15 minutes each in the morning with free refreshments, plus 10 minutes in the afternoon. The work rate steadily increased, from 49.7 to 55.8 finished relays per hour of work.

The women soon realized that the researchers were heavily dependent on their cooperation, and so began to flex their muscles. Most of their requests were trifling: on one occasion the light was too bright for them, on another, they asked for a screen in front of the work bench, so as not to constantly have to see the experimenters looking at them. When this latter measure wasn't immediately implemented, one of the women remarked, 'I'll bet we could get out more work if we had screens in front so we wouldn't have to keep pulling our dresses down.' Shortly thereafter, screens were installed.

The women disliked the medical examinations and the personal questions associated with them. Their attitude didn't change when the doctors came to examine them at the workplace in an attempt to make the relationship more informal. The atmosphere

Source

Gillespie, R. (1991), *Manufacturing Knowledge: A History of the Hawthorne Experiment.* Press Syndicate of the University of Cambridge.

Mayo, E. (1933), *The Human Problems of an Industrial Civilization.* Macmillan.

only improved when the doctors' monthly visits concluded with little parties at which cakes and ice cream were served, while a radio provided the entertainment. Because tea was also served on these occasions, the test room soon came to be referred to simply as the 'T-room' by the other employees at the factory.

When the women started to chat among themselves while they worked, they were upbraided by the researchers. The scientists were afraid that this might lead to falsification of the data. Two of the women, Adeline Bogatowicz and Irene Rybacki, refused to be cowed. Hibarger drew Rybacki's attention to her poor work rate, but she simply responded, 'They tell you to work how you feel, and when you do it, it isn't good enough.'

In fact, from a scientific point of view, it was grotesque to reproach the women for their low productivity, given that it was precisely the aim of the studies to find out what the relationship was between work breaks and productivity.

The situation escalated when Bogatowicz's forthcoming wedding gave her and Rybacki inexhaustible scope for gossip. On 25 January 1928 Bogatowicz and Rybacki were replaced by two other women from the factory.

After the apparent success that the breaks had had in the T-room, they were introduced throughout the factory. However, the researchers still weren't sure whether it was really just the breaks that were responsible for a 25 per cent increase in productivity. In 1928 the people at Hawthorne called in two academics to help them out: Clair Turner from the Massachusetts Institute of Technology, and Elton Mayo from the Harvard Business School.

However, the two professors initially only succeeded in causing confusion. They had the test subjects fill in personality tests, quizzed the women about their eating habits, and measured their blood pressure. No connection was found to exist between these pieces of data and their work rate. Neither did their menstrual cycle appear to have any influence.

In the meantime, the T-room experiments were approaching their 12th phase, in which all the breaks introduced hitherto were rescinded. Even so, the work rate continued to increase. It now stood 19 per cent higher than in phase 3, in which there had also been no breaks.

In the official publication, phase 12 is presented as the period in which the scientists suddenly gained enlightenment. In actual fact, it was a long, steady process that led them to their ultimate conclusions that female workers are emotionally sensitive individuals, who needed to be treated accordingly, and that they form informal groups that compete with one another and consciously control the work rate.

The women themselves attributed their higher output to the less formal atmosphere. This idea had also occurred to the experimenters. Consequently, at the start of 1930, they removed Hibarger, who had a friendly relationship with the women, from the T-room for a spell. However, the experiments that were conducted thereafter showed no clear trend. Further tests outside the T-room on the effect of the pay scale also failed to provide any unequivocal answers as to where the growth in productivity was coming from.

Their last resort was anthropology. A researcher from this discipline was placed in the T-room and instructed to observe the women workers and the experimenters simultaneously, like some field anthropologist studying the natives on a South Sea island. But Hibarger, who had studied and eavesdropped on the women for years, took exception to being made a research subject himself, and so in February 1933 the T-room closed for good.

It is still the subject of controversy today as to why production at Hawthorne in the five years the experiments were running increased by a total of 46 per cent. The data are open to all kinds of interpretations. Was it the pay scale, which was geared to a small group? Or maybe the breaks? Or the fact that just one type of relay was being manufactured in the T-room? Or the better attitude to work that prevailed there? Or the more relaxed and friendly supervisors? Social relationships among the workers are also frequently cited as a factor. For instance, the women made a conscious effort to control their productivity as a group. They realized that it would have been foolish of them to work flat out. If the output were too high, the firm would simply have responded by paying a lower wage for each finished relay. The trick, then, was to carefully regulate their rate of production.

The research team set out with the intention of studying women factory workers, but what they actually encountered were social beings. The practical upshot of the Hawthorne experiments was that concepts such as the work atmosphere, motivation, individual responsibility and identification became the buzzwords for a new generation of managers. Yet the results of the tests were also romanticized. It was claimed that productivity had increased because the women were allowed 'to develop their own values and targets. The test conditions enabled them to quite openly establish norms for their social behaviour in the workplace, and these norms, which no one attempted to interfere with, endowed their work with a lasting significance.' One can only speculate what Adeline Bogatowicz and Irene Rybacki, who were thrown out of the T-room, would have had to say about this, given that the norms they established for themselves evidently weren't ones that suited the supervisor.

Incidentally, the difficulties encountered in this experiment have survived as something of a byword: whenever unexpected influences crop up that have been generated by the experiment itself, people talk about the 'Hawthorne effect'.

"They tell you to work how you feel, and when you do it, it isn't good enough."

1927 Kissing the Culture Medium

At the beginning of the 20th century, people were becoming increasingly aware of infectious diseases. One of the unconventional measures taken against the threat of contagion was the founding of so-called 'anti-kissing leagues'. These societies opposed excessive kissing of children by adults, kisses exchanged between women – or even, like the Paris Anti-Kissing League, against kissing in general.

The French contended that 40,000 germs were transferred in every kiss. They reckoned that if people just stopped and thought about this every time they were about to kiss someone, then it wouldn't be long before the practice died out. They also posed the rhetorical question as to why scenes involving kissing were cut from American and European feature films before they were shown in Japan. It was clearly the case, they claimed, that the Japanese didn't want to fall ill. Or, at the very least, that they had no desire to learn the art of kissing.

However, the Americans weren't about to let some Frenchies ban them from kissing. To this end, the American popular science magazine *Science*

Source

Kraus, J.H. (1927), '40,000 Germs in a Kiss'. *Science and Invention* 15 (169), p. 14.

and Invention conducted an experiment in March 1927. The editors invited a number of men and women to kiss a sterile culture medium in a Petri dish. The culture medium was then placed in an incubator at 37.5°C for 24 hours. During this period, germs that had stuck to the medium during kissing multiplied into small colonies of bacteria that could be seen with the naked eye. By counting them, the scientists could calculate the number of germs that had originally been present but that couldn't be detected individually.

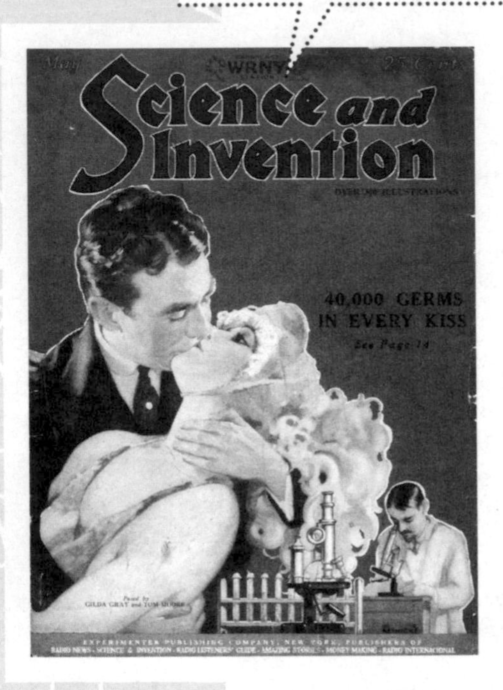

The popular science magazine *Science and Invention* initiated an experiment on kissing.

The laboratory that was charged with this task found on average not 40,000 but just 500 germs, while women who wore lipstick carried around 200 more. *Science and Invention* concluded that this finding finally gave men a scientifically valid reason for refusing to kiss lips that had lipstick on them.

1928 The Lust Curve

The cardiotachometer invented by the American physician Ernst P. Boas was every heart specialist's dream. It allowed automatic and continual monitoring of a person's heart rate during physical activity. All such instruments hitherto had required that they lie still.

Boas and his colleague Ernst F. Goldschmidt immediately set about measuring the pulse of 51 men and 52 women as they went about their daily lives. In the process, they determined the maximum heart rate during various activities: eating (102), telephoning (106), washing in the morning (106.7), listening to music (107.5), dancing (130.6) and gymnastics (142.6). But the front runner, with 148.5 beats per minute, was having an orgasm. Boas and Goldschmidt's book *The Heart Rate* isn't very forthcoming about precisely how this measurement was obtained. 'We were fortunate in obtaining a record of the heart rates of a man and wife during intercourse,' they wrote, before going on to focus on the results. It wasn't particularly earth-shattering to discover that an orgasm makes more demands on the heart than gymnastics, but the two physicians then proceeded to gloss over another truly remarkable feature of the resulting cardiogram as though it were the most

Boas and Goldschmidt's graph showing pulse rates during sexual intercourse. Between 11.25 and 11.45 the female test subject had four orgasms. One researcher ascribed this to the man's 'developed technique'.

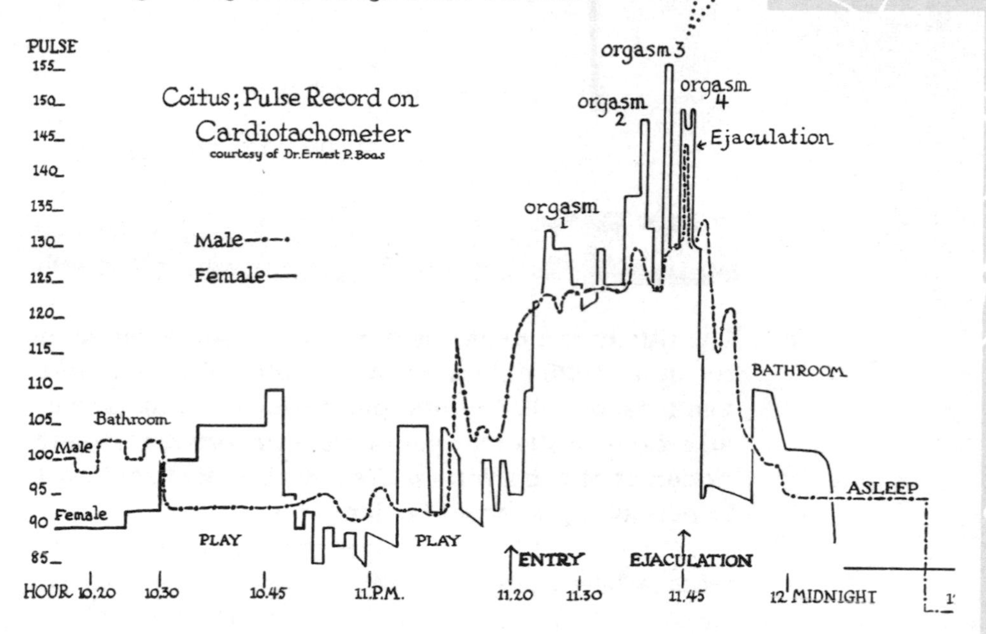

mundane occurrence: 'It shows four peaks of heart rate for the woman, each peak representing an orgasm.' On the evening when the study was conducted, between 11.25 and 11.45 p.m. the woman experienced no fewer than four orgasms – and, what's more, she achieved this with two electrodes attached to her chest with uncomfortable rubber bands and connected to the recording device with more than 100 ft (30 m) of cable.

Boas and Goldschmidt's only comment was: 'The curve of heart rate clearly indicates the strain placed on the cardiovascular system, and helps to explain some cases of sudden death during and after coitus. Pussepp [another researcher] has demonstrated marked rises in blood pressure in dogs during coitus.'

Robert Latou Dickinson, a researcher into sexual behaviour who published the diagram in his 1933 book *Human Sex Anatomy*, was the first to really take note of the four peaks. Yet he attributed them primarily to

Source

Boas, E.P., and Goldschmidt, E.F. (1932), *The Heart Rate*. C.C. Thomas.

the husband's prowess, whom he claimed had a 'developed technique' that allowed him to remain within the vagina for 25 minutes 'awaiting complete satisfaction on her part'.

The fact that the women who took part in orgasm experiments weren't necessarily your average specimens was demonstrated 22 years later by a female test participant whose bodily functions were studied during sexual intercourse (see p. 126).

"awaiting complete satisfaction on her part"

1928 Mamba in the Blood

The title of the paper seemed innocuous enough: in the June 1928 edition of the *Bulletin of the Antivenin Institute*, one 'F. Eigenberger, Physician' published an article entitled 'Some Clinical Observations on the Action of Mamba Venom'. Yet this study almost cost Friedrich Eigenberger his life.

In the spring of 1928 Eigenberger diluted a drop of mamba venom with ten drops of a solution of cooking salt and injected 0.2 cc of this mixture into his left forearm. He then got into his car. – Now, you might well ask why on earth a man would willingly dose himself with the venom of one of the most deadly snakes in the world. And, for heaven's sake, what was he thinking of driving a car after doing so?

Eigenberger was born in Austria in 1893 and emigrated to the USA in 1922, where he took up a post at the Sheboygan Clinic in Sheboygan, Wisconsin. He went on extensive globetrotting expeditions with his wife, giving an account of his travels in public lectures. Without a doubt, the pièce de résistance in his collection of souvenirs – even mentioned in his obituary – was the shrunken head of a Melanesian chief.

The Eigenbergers lived in a striking house built in the Mexican style. The garden was home not only to an impressive orchid collection, but also to a cougar, a leopard and a gibbon. Presumably, the Eigenbergers also kept a green mamba; in any event, the article notes that the venom was 'freshly extracted' just before the experiment.

Eigenberger mentions that he had earlier experimented on guinea pigs, and also conducted tests on himself with rattlesnake venom, suffering painful yet localized swellings. No doubt he expected a similar outcome from his experiment with mamba venom. But things panned out differently. After administering the venom, Eigenberger suddenly found himself becoming hypersensitive to all external stimuli: 'The vibration and motor noise of my car appeared so loud and annoying that I thought all four tyres were flat and it made me stop and look before I realized the reason.'

After 20 minutes he experienced a mild sensation of having been poisoned, but shortly thereafter began to feel desperately ill. In an attempt to lessen the effect of the poison, he applied a tourniquet above his elbow, lanced the swelling – which had spread over an area of his arm 4 inches (10 cm) long – and poured hot potassium permanganate solution over the bleeding wound in the hope of washing out the venom.

But by this time the toxin had spread throughout his entire body. Eigenberger experienced 'a numb feeling around the lips and chin and in the tip of the tongue, spreading rapidly over the entire face and down the throat'. He also lost all sensation in his fingers and toes. His eyes were sore, and he found it hard speaking or swallowing. 'The general feeling was extremely bad, however I continued to walk back and forth, as I felt I would faint if I lay down.' When his pulse reached 160, Eigenberger urgently requested an injection of strychnine – a somewhat curious demand from a modern perspective, given that we now know strychnine to be a stimulant. But perhaps Eigenberger believed in the reputation it had at that time as an antidote to snake bites. Six hours later, his whole body reacted painfully to the slightest touch. And yet the next day, after a night in which he suffered influenza-like symptoms, the effects of the poison began to abate.

Eigenberger was well aware that his own miscalculation had almost caused his death. The reason why he had injected such a large amount of mamba venom was his observation that the venom had a relatively slow effect on mice. They usually survived longer after being bitten by a mamba than they did after a rattlesnake bite. Yet the two venoms work in entirely different ways. Rattlesnake venom targets the blood vessels and the blood cells. It can destroy tissue and either delay or accelerate

Source

Eigenberger, F. (1928), 'Some Clinical Observations on the Action of Mamba Venom'. *Bulletin of the Antivenin Institute of America* 2 (2), pp. 45–46.

blood clotting. But in many cases it results only in the painful swellings that Eigenberger described. By contrast, the venom of the mamba is a neurotoxin that affects the central nervous system and can paralyse the respiratory system and heart.

In his paper, Eigenberger fails to address either the question of why he carried out the experiment in the first place or why he drove off in his car while his bloodstream was full of snake venom. Nevertheless, he seems not to have undertaken any more experiments of this kind on himself up to his death in 1961. Evidently the risk was too great even for an eccentric pathologist with big cats in his garden and a shrunken head on his desk.

1928 The Living Dog's Head

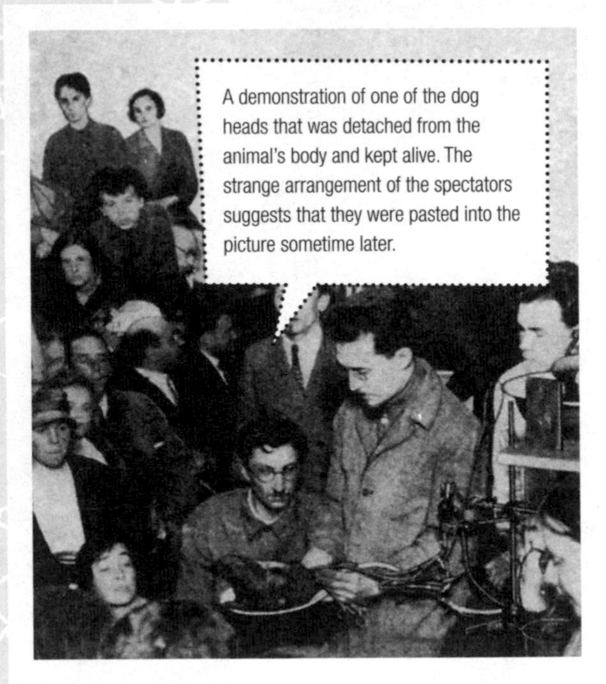

A demonstration of one of the dog heads that was detached from the animal's body and kept alive. The strange arrangement of the spectators suggests that they were pasted into the picture sometime later.

In the photographs, it looks like some kind of circus act. In the middle of the picture is the dish containing the severed head of the dog, from which several tubes lead to a rack holding a pump, a bottle and a dish full to the brim with blood. All around, tightly packed, stands a crowd of curious onlookers, who are about to bear witness to a scientific miracle: the dog's head is alive.

The Russian surgeons Sergei Brukhonenko and S. Tchetchuline had removed the dog's head from its body in an operation that the popular science magazine *Science and Invention* informed its readers 'might appear cruel and inhumane' – but then went on in the very next sentence to point out the great benefits of animal research. Now the dog's head lay there with its mouth half open, and it seemed as though the scientists felt obliged to prove to the assembled laymen that the head really was still alive. And so they shone a torch into its eyes until its pupils

Source

Brukhonenko, S.S., and Tchetchuline, S. (1929), 'Expériences avec la tête isolée du chien' ('Experiments with the severed dog's head'). *Journal de physiologie et de pathologie générale* 27 (1), pp. 31–45, 64–79.

contracted, smeared vinegar around its mouth, which it promptly licked off, caused its eyes to well up with tears by making it sniff bitter quinine, and fed it sweets, which promptly emerged from the stump of its oesophagus after it had swallowed them.

Is this what eternal life looks like? A living dog's head in one of the Russian experiments.

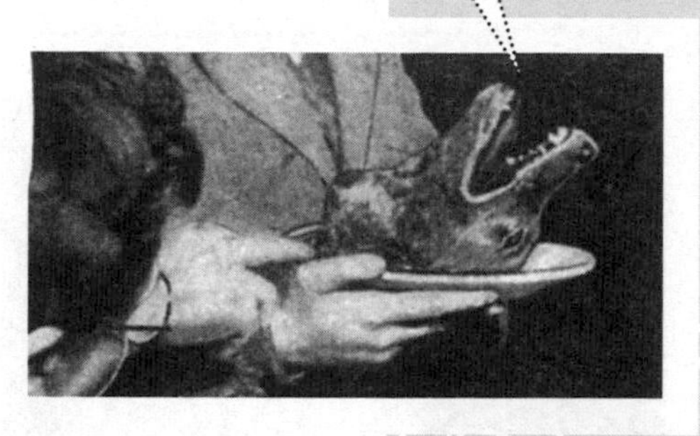

Brukhonenko and Tchetchuline were not the first people to conduct experiments with severed heads (see page 41), but unlike earlier attempts, they kept the head alive by means of a mechanical heart. Blood was channelled from the carotid arteries through rubber tubes into an open basin, where it was mixed with oxygen, and from there into a bottle placed a little higher than the dog's head; from the base of the bottle the blood flowed back under constant pressure into the carotid arteries. The blood was chemically treated beforehand to prevent it from clotting.

madsciencebook.com A documentary film by the Soviet Film Agency containing original footage of the experiment. Although many people have called the authenticity of this film into question, experts are in no doubt that it is genuine.

This bizarre experiment fired people's imaginations. Might it also be possible with a human head? Was this what eternal life on Earth would look like? A French researcher proposed founding a Society for the Avoidance of Death, while *Science and Invention* enthused: 'Do not even the widest imaginations of our modern science-fiction writers pale into insignificance because of the steady advance and progress of scientific research?'

1930 Mr Skinner's Box

Burrhus Frederic Skinner couldn't have known that the box that he knocked up in the workshop in the Psychology Department at Harvard University would later become one of the most famous pieces of apparatus ever assembled for a scientific experiment. The Skinner box went on to feature in numerous cartoons, and it was parodied in the TV cartoon series *The Simpsons*. Even a rock band called itself 'The Skinner Box'. A connection was even drawn between this cage and the supposed suicide of Skinner's own daughter.

madsciencebook.com A short documentary on Skinner's work.

Skinner was 26 years old when he began his search for an instrument that could monitor the behaviour of rats. The maze that was currently popular with researchers (see p. 55) didn't

The psychologist B.F. Skinner with a Skinner box, a cage constructed for the study of learning behaviour in animals.

strike him as ideal. As he later wrote in his memoirs: 'The animals' behaviour was composed of too many different "reflexes" and should be taken apart for analysis.' And so he concentrated on just one small part of the test circuit: a soundproofed box with a noiseless door, from which a rat could be released into a maze without causing disruption. But he soon dropped the maze part of the apparatus altogether. Using clockwork-like measuring instruments, he tried to determine the animals' movements. Yet the findings were too haphazard to be properly evaluated. Skinner read Pavlov, who 30 years before had discovered classic conditioning (see p. 63). This technique enabled innate reactions to be linked to new stimuli. However, Skinner didn't just want to investigate existing reactions but to find out how new behaviours arose.

He finally hit on the idea of equipping his experimentation box with a lever. Whenever the rat pressed down on this, it received a pellet of dried food. Of course, the rat didn't know this from the outset and only triggered feeding when it happened to touch the lever by accident. But after scoring several such lucky strikes, it appeared to have learned the connection, and consequently the time that elapsed between pushes on the lever became ever shorter. Skinner had thereby discovered a simple yardstick for measuring a rat's changes in behaviour: namely, the frequency with which a particular behaviour occurred.

Unlike Pavlov's dogs, the animal in Skinner's experiment wasn't exhibiting an innate reaction, but was learning a new behaviour. The theory that Skinner developed on the basis of this comprises three elements: first, living creatures constantly exhibit spontaneous behaviour; second, the consequences of a particular behaviour – positive or negative – diminish or increase the chances that an organism will repeat that behaviour; and third, it is the environment that determines these consequences.

He called the whole process 'operant conditioning' (to distinguish it from Pavlov's 'classical conditioning').

Skinner had no interest in what went on in the brain in this process. Because it was impossible to directly observe intellectual activity at work, he considered it unscientific to engage with it at all. Together with John B. Watson (see p. 77) he was a leading figure in the movement known as behaviourism, which views animal and human behaviour exclusively as a series of reactions to external stimuli.

'Oh, not so bad. The light goes on, I press the lever, and they write me a cheque. And how are things with you?'

The Skinner box had one great advantage over earlier pieces of equipment such as the maze: after the rat had pressed the lever and got the food, everything was then automatically ready for the animal's next action without any human intervention. An automatic writing device recorded the exact time of each depression of the lever, and from this data Skinnner could study the animal's learning behaviour under various different conditions. For instance, what happened when the rat had to press down five times in succession before it received a food pellet, or alternatively if it was only rewarded after a random number of depressions? What if it could avoid punishment by carrying out a particular action? And how could a learned behaviour be erased once more? One could almost say that the Skinner box automated research into animal behaviour.

Another cartoonist's take on Skinner's operant conditioning.

On the face of it, the method behind operant conditioning appears banal – reward reinforces a particular behaviour, while punishment discourages it. Yet Skinner used it to teach animals far more than simply how to press a lever. For example, he taught a pigeon how to play a tune on a toy piano, and two pigeons how to play table tennis. The trick in all this was not to reward the animals only for attaining the overall goal, but for every small interim step. And so the pigeon was given grain when it just happened by chance to hit the correct first note with its beak on the toy piano in the Skinner box – and again when it got the second note right, and the third, and so on until it could play the entire children's song 'Over the Fence is Out, Boys'.

Humans could use this kind of training to adapt animal intelligence to a whole variety of tasks. During the Second World War, Skinner worked for the US military on a highly unusual anti-shipping bomb-guidance system, involving conditioned pigeons placed in the nose cone of the projectile. Depending upon the position of the target ship, which they could see through ports in the nose of the bomb, they would tap with their beaks at different points on a screen. These signals were then used to guide the bomb. This guidance system worked fine in the laboratory, but not in the field.

Skinner didn't actually coin the term 'Skinner box', but it quickly became popular. He was even suspected of having raised his second daughter, Deborah, in a Skinner box, and subsequently a rumour began to do the rounds that Deborah had ended up in a psychiatric institution and taken her own life. The seeds of this urban myth were sown by the October 1945 edition of the *Ladies' Home Journal*. This women's magazine ran an article on the soundproofed heated crèche that Skinner had built for Deborah. Unfortunately, the title of the piece was 'Baby in a Box', which misled many readers into thinking that Deborah had been put in a Skinner box where, like her father's rats and pigeons, she was forced to take part in experiments. Nowadays Skinner's daughter, who works as an artist in London, surfaces from time to time in the press to scotch the persistent rumours that she committed suicide.

Skinner was a controversial figure in American academia. His findings were particularly influential in education, since there were clear parallels between his experiments and the way teachers encouraged or rebuked their pupils. As far as Skinner himself was concerned, the world was one great Skinner box, and he was adamant that it could be used to explain the whole gamut of human behaviour. In his controversial book *Beyond Freedom and Dignity* (1971), he proposed the widespread introduction of conditioning techniques for the good of humankind, as a method of training people to act in ways that were beneficial to society.

Source

Skinner, B.F. (1938), *The Behaviour of Organisms: An Experimental Analysis.* Appleton-Century.

1930 Chinese Travelling Companions

Richard T. LaPiere could have guessed the answer the minute he placed the call. Would the hotel be prepared, he enquired, to put up 'an important Chinese gentleman'? 'No,' said the person at the other end of the line.

Yet, two months before, this professor of sociology from Stanford University had spent a night in that very hotel in the company of a Chinese couple who were friends of his. The establishment in question was the best hotel in a small town that was notorious for discriminating against people of Asian origin. But, to LaPiere's surprise, they had got rooms without any bother.

The hotel manager had said one thing on the phone, but done the exact opposite two months earlier. Was this just an isolated case? The act of an indecisive personality? Or was there more to it? An experiment would surely sort the wheat from the chaff. The question was an important one: did people have a fundamental problem in telling others what they would do in a given situation? The social sciences were based in large part on studies involving questionnaires: Do you believe in God? Would you give up your seat on a tram to an Armenian woman? What do you think of Asian people? In evaluating the responses, researchers made the tacit assumption that people would also act accordingly in everyday situations. But if this assumption weren't true, then many of the results were either invalid or at the very least meaningless. After all, the idea was to find out what people would really do, not what they said they would do on paper.

So, in 1930–1, LaPiere undertook two extensive trips around the USA in the company of his Chinese friends. Travelling by car with the young couple, he covered some 10,000 miles. They stayed in a total of 66 hotels and ate in 184 restaurants. Only once were they turned away, when the owner of a low-rent bungalow park peered into the car and responded to LaPiere's inquiry whether he had a chalet free with a curt 'No, I don't like Japs.'

Otherwise, the three travellers were treated with the utmost civility. Many people in the countryside had never come across Chinese people before, but the stir that their presence created manifested itself not as rejection but on the contrary as extreme courtesy.

In the course of a trip across the USA, the sociologist Richard T. LaPiere established that people's convictions aren't necessarily a reliable indicator of their actions.

LaPiere kept a meticulous record of all the encounters they had with receptionists, porters, bellhops and waitresses. Of course, these observations were subjective, as he himself was quick to concede, but when all was said and done these weren't laboratory experiments in which every last element could be controlled.

So as to keep his own influence on proceedings to a minimum, whenever possible he left it to his Chinese friends to ask for the rooms while he attended to the baggage. He would often send them off to restaurants on their own and join them later. To ensure that they acted completely naturally, he hadn't even let on to them that they were taking part in an experiment.

From his notes at the end of the trip, LaPiere concluded that race wasn't the main factor influencing people's attitudes, but rather smart clothing, a timely smile, and his friends' perfect command of English. Summing up the experiment in his now-famous essay 'Attitudes vs. Actions', he wrote: 'A Chinese companion is to be recommended to the White travelling in his native land.'

But how did this square with people's attitudes as expressed in questionnaires? LaPiere knew from surveys that Americans were strongly prejudiced against Asians. In order to have some basis for comparing people's actual experiences and their attitudes, six months after the trip – and without revealing his identity – he sent all the hotels and restaurants they had visited a letter with the question 'Would you accept members of the Chinese race as guests in your establishment?' Out of the 128 replies he received, just one answered 'Yes'. Practically all the others responded by saying that they would turn Chinese people away. One or two couldn't decide one way or the other.

LaPiere immediately fell to wondering whether his trip itself might actually have caused this negative outcome. Perhaps his sojourn with his Chinese friends at these establishments had left their owners with a bad impression – not that he had noticed any negative reactions at the time. To test this hypothesis, he sent the same letter to various hotels and restaurants that they had not visited along the route. The result was just the same: nobody wanted anything to do with Chinese people.

'On the basis of the above data, it would be foolhardy for a Chinese to attempt to travel in the United States,' he wrote. But experience had revealed a different picture. The sociologist concluded from this that questionnaires are fundamentally flawed in predicting how people will act in any given situation: 'A questionnaire will reveal what Mr A writes or says when confronted with a certain combination of words. But not what he will do when he meets Mr B. Mr B is a great deal more than a series of words. He is a man and he acts.'

Source

LaPiere, R.T. (1934), 'Attitudes vs. Actions'. *Social Forces* 13, pp. 230–23.

1931 An Ape for a Kid Sister

Of all the unusual childhood experiences ever undergone by an ape, that of Gua the chimpanzee must rank as the most bizarre. On 26 June 1931, when Gua was just seven months old, she was sent to live with a human family. Not as a pet, but as a full member of the family, one who was treated exactly the same as their ten-month-old son Donald.

Winthrop Kellogg was 29 years old when, in 1927, he first got the idea for his unorthodox experiment. In all likelihood, the catalyst was an article about two small girls who had been found living in a cave in eastern India with a pack of wolves. They ate and drank like wolves, and only used their hands to crawl on all fours. Although, following their discovery, they did learn to walk upright, their carers could never get them to stop howling at night or pouncing on birds and eating them raw. Also, they scarcely learned to speak.

Experts ascribed these shortcomings to the wolf-children's lack of intelligence. Yet Kellogg disagreed: the children had learned their wild behaviour from living with the wolves, and would have found it impossible to adapt to their new environment because it is extremely difficult to unlearn behaviours imprinted in early childhood.

Donald and Gua spent nine months together.

In order to test this hypothesis, wrote Kellogg, all that one needed to do was to turn a normal, averagely intelligent baby out into the wilderness and study its behaviour. Yet for all his 'scientific enthusiasm' for this plan, it was of course wholly impracticable on ethical and legal grounds. But no such qualms prevented the converse experiment from taking place – namely, having a young ape grow up like a human baby.

The adoptive parents of the baby ape were under no circumstances ever to treat it like an ape. It was to be kissed, made a fuss of and pushed around in a pram, and taught how to eat with a spoon and use a potty. Initially Kellogg provided children from a local crèche who had 'understanding parents' as playmates for the chimp. Yet he realized that it would be even better if the ape was adopted by parents who had a child of their own. This would allow a direct comparison to be made of the baby's and chimp's development.

In this way, Kellogg hoped to clarify once and for all whether nature or nurture – in other words, environmental or hereditary factors – had the upper hand in a child's development. If the ape failed to develop in the same way as the child, then this would mean that the animal's inherited instincts were dominant. But if, on the other

Gua was better than Donald at many things, except for one key aptitude: the child was more skilful than the chimpanzee at mimicry.

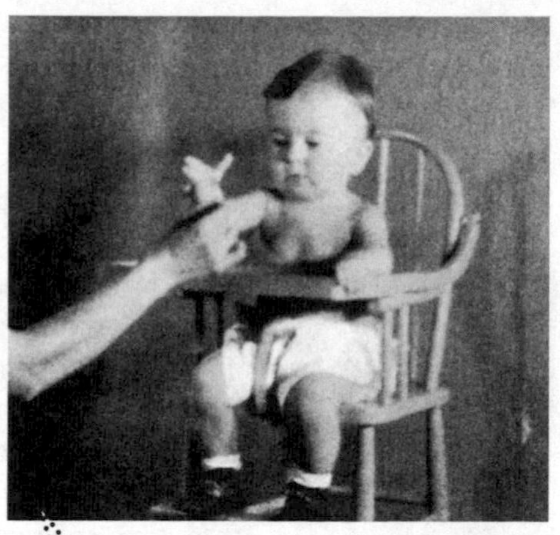

Which is the stronger – nature or nurture? Could Gua be brought up to act like a human being? The ape and the human child were treated exactly the same.

hand, the ape showed typically childlike reactions, this would be evidence of the power of the environment.

But before the experiment could begin, Kellogg had to convince his wife Luella to take part. For the adoptive parents he had chosen were none other than himself and his spouse, with the control being the child that they planned to conceive. One passage in the foreword to the book *The Ape and the Child*, in which he describes the experiment, hints at the fact that it took place against Luella's wishes: 'Indeed, the enthusiasm of one of us met with so much resistance from the other that it appeared likely we could never come to an agreement upon whether or not we should even attempt such an undertaking.' However, Winthrop Kellogg finally prevailed. In the interim, he had been appointed professor of psychology at Indiana University. For the duration of the experiment, however, the family lived near the Yale Anthropoid Experiment Station at Orange Park, Florida.

From the moment that Gua arrived, Luella and Winthrop Kellogg devoted themselves entirely to the experiment. Morning, noon and night they were scrupulously fair about treating Gua and Donald the same. They weighed them both daily and measured their blood pressure and body mass. They tested their visual perception and motor skills. To test their susceptibility to fear, Kellogg fired a blank pistol behind their backs and filmed their reactions.

Kellogg's description of the experiment indicates that he was a stickler for scientific accuracy: 'The differences between the

Luella Kellogg pushing Donald and Gua in a pram. At the outset, the experimenter's wife had deep misgivings about allowing the experiment to be conducted with her own son.

skulls can be audibly detected by tapping them with the bowl of a spoon or with some similar object. The sound made by Donald's head during the early months is somewhat in the nature of a dull thud, while that obtained from Gua's is harsher, like the crack of a mallet upon a wooden croquet or bowling ball.'

The Ape and the Child is a painstaking account of Gua and Donald's development. This makes it all the more surprising why no explanation is given as to why the experiment was terminated after nine months. The psychologist Ludy T. Benjamin, who quizzed former students of Kellogg on the matter, suspects that things took an unexpected turn. Certainly, Gua showed an astonishing ability to adapt to her human surroundings, obeying commands better than Donald, asking for forgiveness by giving people kisses, and making it known from an early stage when she had to go to the toilet. She also grasped more quickly than Donald that she had to clamber on a chair to reach a biscuit suspended from the ceiling. Yet Donald had the advantage over her in one aspect: he was better at imitating. Gua was the leader, who discovered toys and invented games, while Donald copied what she did. The same went for language skills. Donald would imitate Gua's call for food to perfection and would use the same panting noises as her to ask for an orange.

At the age of 19 months, when the experiment came to an end, Donald could say just three words, whereas an average American child of that age has a vocabulary of around 50 words and has begun to use them to form sentences. In other words, Winthrop Kellogg planned to bring up an ape to be a person and ended up teaching a person to be an ape. We might reasonably assume that Luella was no longer willing to stand idly by and watch this happen. Gua was returned to the anthropoid colony at Orange Park after the experiment. She had difficulty adjusting to her caged existence with her original mother and died the following year.

The experiment caused quite a stir, and Kellogg came in for some harsh criticism. Many people considered it irresponsible of him to subject a child to such a procedure, and Kellogg was accused of courting sensation and publicity. He himself later conceded that this type of research demanded a 'determined scientist' capable of facing down those who lampooned the experiment because

Source

Kellogg, W.N., and Kellogg, L.A. (1933), *The Ape and the Child.* Hafner Publishing Company.

they fundamentally misunderstood its aims.

After publication of *The Ape and the Child*, Winthrop Kellogg turned his attention to other areas of research. He died in Florida on 22 June 1972 at the age of 74, his wife Luella following him one month later.

The psychologist Winthrop Kellogg carrying out a test to ascertain comparative susceptibility to fright. The children and Gua were caught on film as he unexpectedly fired a shot behind their backs.

Donald Kellogg quickly made up for lost time in his speech development and later went on to study medicine at Harvard Medical School. He eventually became a psychiatrist. A few months after the death of his parents, he took his own life, a fact that is often glossed over in descriptions of the experiment. 'Some people, of course, try to link his suicide to the early experiences with Gua and his separation from her, but a much more parsimonious explanation for his depressions is that he was raised by a father who could be extremely harsh and was demanding of perfection in all with whom he interacted,' wrote the historian of psychology Ludy T. Benjamin.

Donald Kellogg's son Jeff was nine years old when his father killed himself. Unlike Benjamin, he is convinced that his father's suicide was a direct result of the experiment: 'As Ludy mentions, his [i.e. Donald's] father's treatment was not empathetic or supportive, but more strictly performance-oriented. Still, Ludy dismisses the rather obvious contribution from the experiment, perhaps because he is a research psychologist.' In an unpublished monograph on the aftermath of the experiment, he calls the suicide a 'forty-five year murder'. (It was forty-five years prior to Donald's death – in other words, before he was born – that his father Winthrop began planning the experiment.)

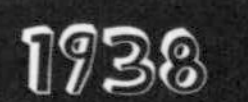

1938 The 28-Hour Day

Nathaniel Kleitman had conducted a fair few extraordinary experiments over the years, but this was the first time that he'd been blinded by floodlights at the end of one. As he and his student assistant Bruce Richardson emerged from Mammoth Cave on 6 July 1938, the massed ranks of film crews

and photographers were waiting at the entrance for them. The next day's papers printed pictures showing two miserable-looking specimens, whose full beards, long coats and wet anorak hoods made them look like tramps. These two researchers from the University of Chicago had spent the preceding 32 days in one of the cave's chambers trying to fathom the mystery of sleep.

Kleitman, 43 at the time, was used to experimenting on himself. On one occasion he had played the guinea pig to find out the effects on a human of 180 hours of sleep deprivation, while another time he had tried unsuccessfully to acclimatize his sleep–wake rhythm from the normal 24-hour cycle to one lasting 48 hours. This involved him spending a month staying awake for 39-hour stretches and then sleeping for 9 hours. Meanwhile, one of his students tried shifting to a 12-hour cycle, sleeping for two periods of 3½ hours per day – between 4.00 and 7.30 in the morning and again between 4.00 and 7.30 in the evening. This attempt to reset sleeping patterns also failed.

One of the great unsolved mysteries of sleep research at that time was whether the human sleep–wake rhythm of 24 hours was merely a habit – geared for practical purposes to the length of the day but changeable at any time – or whether people had an internal, hard-wired body clock.

Two sleep researchers spent 32 days in this chamber in Mammoth Cave. The extended legs of the beds were stood in buckets to guard against rats.

Clearly, from his experiments, Kleitman had established that human sleep rhythms couldn't be doubled or halved, so for his next experiment, at the University of Chicago, he opted for two rhythms that were closer to the natural 24 hours. He settled on 21 and 28 hours, since a normal seven-day week can be divided exactly into either eight 21-hour days or six 28-hour days. This also enabled the two test subjects – one of whom was Kleitman – to continue their work at the university even while the experiment was running.

Kleitman determined whether a test subject had become acclimatized to a changed sleep pattern by measuring their body temperature, which customarily falls when a person is sleeping because their metabolism slows down, and reaches its peak when they are awake. If the rise and fall in temperature were to change in accordance with the new periods of sleep and wakefulness, then by definition the body would have acclimatized to the new rhythm.

Source

Kleitman, N. (1963), *Sleep and Wakefulness*. University of Chicago Press, pp. 175–182.

However, the results of the 21/28-hour tests at the university were inconclusive: although the temperature rhythm of one of the students taking part in the experiment had adjusted to the new conditions, Kleitman's rhythm remained around the 24-hour mark.

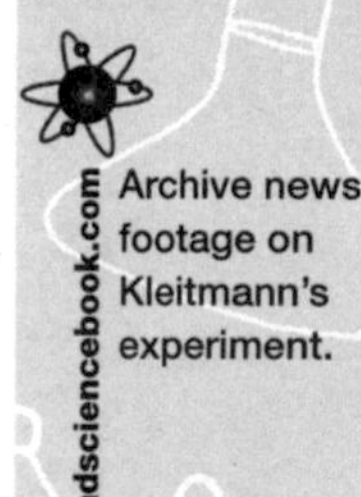

Archive news footage on Kleitmann's experiment.

One possible disruptive factor was the place where the experiments were being conducted – maybe the natural rhythm of daylight was preventing adjustment, or possibly the increased noise level and higher temperatures during the day were to blame. Accordingly, Kleitman began to search for a location where there was no difference between day and night. He found it in a rock chamber, 65 ft (20 m) wide and 26 ft (8 m) high, and 130 ft (40 m) under the surface, in the Mammoth Cave in Kentucky. Located just to the side of a passageway known as 'Audubon Avenue', the chamber was a place of constant darkness and quiet. Year in, year out, the temperature remained at exactly 12°C. Was this, then, the ideal place to try the 28-hour day?

The Mammoth Cave Hotel furnished the 'Apartment on Audubon Avenue' – as the underground chamber became know in the press – with a table, chairs, a washstand and two beds on extended legs (to counter both the damp and the rats). The hotel's cook also made daily deliveries of food to the same address.

The plan envisaged that Kleitman and Richardson would sleep

The sleep researchers Nathaniel Kleitman (left) and his student Bruce Richardson at their morning ablutions.

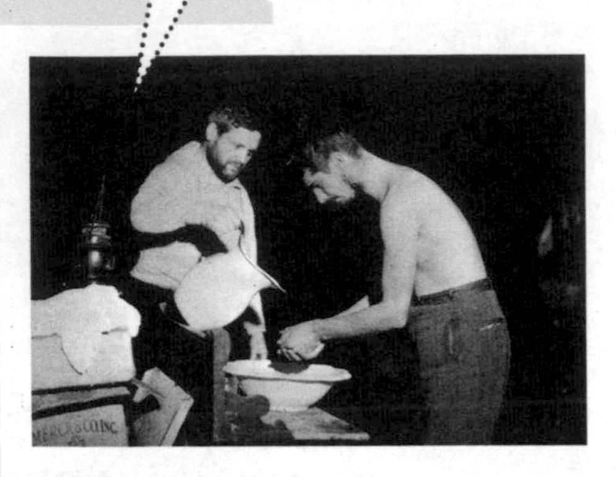

for 9 hours, work for 10 and then have 9 hours' worth of leisure. In the periods when they were awake, the two men took their temperature every 2 hours, and when they were asleep every 4 hours.

After just a week Richardson had adapted to the new cycle: his body temperature took on a 28-hour rhythm. In contrast, Kleitman, who was 20 years older, failed to adapt even by the end of the test period. He always found himself getting tired at 10.00 in the evening and lively again 8 hours later, irrespective of whether the timetable prescribed work, sleep or recreation.

Once again, then, the results were ambiguous. But, as Kleitman told journalists, at least he had ascertained that he was capable of growing a respectable beard.

Later experiments showed that human beings do indeed have an internal clock. For most people, it is set at around 24 hours and is recalibrated on a daily basis by the actual amount of daylight.

1945 The Great Famine

It began four months after the start of the experiment. Up to 6 July 1945, Lester Glick had experienced no problems when visiting restaurants 'to watch people eat'. As the so-called 'buddy rule' required, on that day – just as on all previous days – he wasn't out and about on his own but in the company of Jim, another of the test subjects. Together, they had to watch while a well-dressed woman ordered a pork chop, played around with it for a bit and then left half of it uneaten on her plate. But when she then proceeded to push aside most of a coconut cream tart, that was the last straw for the two men.

After the woman had paid, they followed her, stopped her and gave her a thorough dressing-down about world poverty and how she was contributing to it. The woman yelled at them and ran off. She could not have known that Lester and Jim had, since 12 February, had to survive on two frugal meals a day, consisting chiefly of bread, potatoes, turnips and cabbage.

These two conscientious objectors had responded to an appeal launched by the civilian public service. 'Will You Starve That They Be Better Fed?' ran the slogan on the leaflet that the biologist Ancel Keys had circulated to community-service workers. Keys was the founder of the Laboratory of Physiological Hygiene at the University of Minnesota at St Paul, and during the Second World War worked for the US military, testing the ration packets issued to troops, finding out what kinds of diet contributed to exhaustion, and investigating whether people lost vitamins when they sweated. Towards the end of the war, he got interested in a new question: 'By that time I realized that there were a lot of people, millions of people, who were in semi-starvation conditions. I wanted to find out what would be the effect of that, how long it would last, and what would be required to bring them back to normal.'

More than 100 conscientious objectors volunteered for the experiment, and 36 were selected. On 19 November 1944 they moved into accommodation at the university. For the first three months they were fed a normal diet, while Keys tested their state of health, their average nutritional intake, and other details of their metabolisms. The experiment proper began on 12 February 1945. The test subjects were now given just two meals a day, one at 8.30 a.m. and the other at 5 p.m. The three menus that were rotated over a period of six months corresponded to the food that people were eating in those parts of Europe affected by starvation. These meals provided a total daily energy value of 1500 calories – just half what the men had been consuming previously. Keys calibrated the precise amount of food supplied to the individual body weight of each of the test participants. His aim was to make each of them lose one-quarter of his weight over the six months. The starvation phase was than followed by a three-month period of rehabilitation, in which the test subjects were divided up into groups

Starving for 48 weeks: the participants sunbathing shortly before the end of the experiment. Some of them later became chefs.

A skeleton on a treadmill: the endurance of the test subjects was measured at regular intervals.

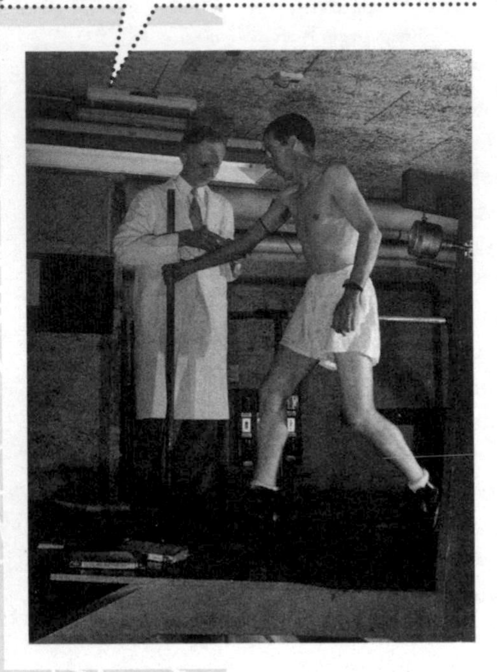

and fattened up once more using various different menus.

Four years after the experiment, Keys published his findings plus all the data he had collected in the 1400 pages of his seminal work *The Biology of Human Starvation*. The experiment had investigated not only physical processes such as weight loss, hair loss, sensitivity to cold and changes in the body chemistry and internal organs, but also the effect of poor nutrition on intelligence, powers of comprehension and personality.

The test subjects were required to work 15 hours a week – in the laboratory, the laundry or the dormitories – and walk a minimum of 20 miles (30 km) in the open air, in addition to spending half an hour on a treadmill. They were allowed to attend regular courses at the university, and had the weekends to themselves.

Some of the most interesting findings from Keys's experiment were those relating to the mental changes that came about as a result of hunger. Many men fell prey to apathy and depression. Sheer hunger made everything else irrelevant. They neglected their personal hygiene and table manners, became reclusive, and could only summon up interest for things that were related to food. Moreover, they lost their sex drive. Romantic films bored them – except for the scenes where people were eating.

On 10 May Lester Glick wrote in his diary: 'My hunger has taken on new dimensions that I could never have imagined. It seems that my bones, my muscles, my stomach and my mind have united in their longing for FOOD!' In common with many of the other men, he spoke less and less, and recipes became his preferred reading matter. Their compulsive fixation on food manifested itself in uncharacteristic behaviour: comparing the prices of provisions in newspaper advertisements, watching other people eat, collecting recipe books and buying cooking implements such as hotplates and teapots. After the experiment, three of the test subjects changed their profession and became chefs.

As the starvation phase drew to a close, some of the men would spend two hours over their meagre meals. They would endlessly rearrange the food on their plates in an attempt to make it seem larger than it actually was. And no sooner had they had licked their plates clean than they began planning the order in which they would eat their next meal.

Initially the men were allowed to have as much coffee and chewing gum as they wanted. But then some of them started drinking 15 or more cups a day and got up to 40 packs of chewing gum in the same period. Accordingly, Keys restricted their daily ration to a maximum of 9 cups of coffee and 2 packets of gum.

Not all the test subjects made it through to the end of the experiment. One of them flipped out in a grocery store, wolfing down several biscuits, a bag of popcorn and two overripe bananas, which he then promptly sicked up. Another stole swedes and candy bars. On one occasion, Lester Glick took the lead out of a pencil and started chewing the wood. He recorded in his diary that 'it tasted all right', and later wrote 'I think about how cannibalism is a terrible option for a starving person, and try to put it out of my mind, but I can't seem to stop thinking about it.'

The temptation to eat in secret was so great that Keys was forced to introduce the 'buddy system' after two months: this prescribed that nobody was permitted to leave the laboratory unless they were accompanied by at least one other person.

Throughout the entire 24 weeks of the experiment, the men longed for the onset of the final phase. And yet, when it came, the rehabilitation phase turned out to be an anticlimax: the size of the portions was only increased gradually and their feeling of constant hunger hardly abated. On 20 September 1945 Glick noted in his diary: 'We're seven weeks into rehabilitation and our starvation symptoms have not abated significantly. Our look, our hunger, our minimal weight gain all verify our minimal rehabilitation.'

On 20 October 1945 at 5.00 p.m. the group took their final communal meal together. This was the first meal for 48 weeks at which no restrictions whatever were imposed. As Keys wrote: 'The urgent desire for dietary freedom expressed by the men was extreme, the postponement for another week could have produced severe emotional crises and possibly open rebellion.' Yet the sumptuous banquet filled up many of the test subjects quicker than they had expected, as Keys explained: 'The end of the meal found most men gazing in unbelief at the food – their

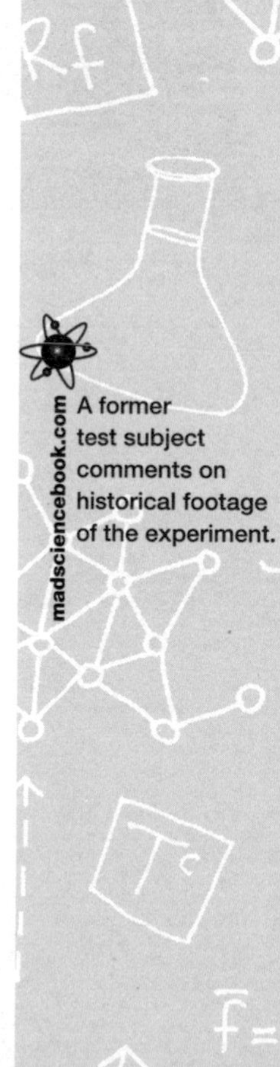

madsciencebook.com

A former test subject comments on historical footage of the experiment.

food – which they could not eat.'

The experiment resulted in no lasting harm to the participants, but it did take months for their bodily functions to normalize. After the experiment, many of the men claimed that they often felt hungry even though they were unable to eat any more. Many of them ate until they were sick, and then immediately began the process all over again.

Because of the similarity of this behaviour to the symptoms of bulimia, Ancel Keys's work now plays a major role in research into eating disorders. The compulsive fixation of the starving test subjects on food, their apathy and their reclusiveness all resemble the behaviour manifested by anorexics. These forms of behaviour are often regarded nowadays as the causes of eating disorders, but it may be the case – as with the starving conscientious objectors – that they are simply the effects of starvation.

For the men involved, the experiment remained 'the most significant event in their lives', and, right up to the 1990s, they regularly held reunions.

1946 A School Dropout Makes It Rain

In all probability, hardly anyone in Pittsfield would have noticed that it snowed that Wednesday. The sparse flakes that fell from a towering cumulus cloud over Massachusetts on 13 November 1946 melted and evaporated before they reached the ground. And even if anyone had looked up and noticed the precipitation falling from the base of the cloud, they couldn't possibly have guessed the significance of this low-key natural spectacle. To do that, our hypothetical observer would have had to make a connection between the snowfall and the light aircraft that kept banking around the cloud.

The occupants of the single-engined Fairchild were the researcher Vincent Schaefer and the pilot Curtis Talbot. They had just climbed through the cloud to a height of around 13,000 ft (4000 m), and Schaefer had tipped 2.4 pounds (1.1 kg) of dry ice out of the window. It looked for all the world as though he were sowing grey seeds the size of walnuts. Nor did he have to wait long for his harvest; snow immediately began to fall from the strip of cloud they had just flown through. 'I turned to Curt and we

The 'cold box' that Vincent Schaefer devised allowed him to create clouds in the laboratory.

shook hands as I said, "We did it!",' Schaefer later wrote in his laboratory diary.

It seemed as though one of man's oldest dreams had just come true. No more superstitious casting of spells was required, no rain dances, no fervent prayers to the gods. As one of Schaefer's colleagues put it following the successful flight: 'Well, Schaefer made it snow this afternoon over Pittsfield! Next week he walks on the water.'

The following day the whole world learned about Schafer's experiment, when the *New York Times* ran an article with the headline 'Three-Mile Cloud Made into Snow'. Meanwhile, the *Berkshire Evening Eagle* published a piece on Schaefer himself: 'The man who made it snow over Graylock dropped out of school early on.' Indeed, Schaefer had never graduated from high school. He acquired his encyclopaedic knowledge of chemistry and physics during his long employment with the firm General Electric, under whose auspices he also conducted the experiments. The head of the GE laboratory, Irving Langmuir, was bullish about the future of weather modification, firmly believing that the technique of seeding clouds with dry ice could be used to 'steer heavy snowfalls away from city areas and provide snow for winter resorts'.

Langmuir had already won the Nobel Prize for Chemistry for the chance discovery of certain unexpected properties in the action of rain and snow. During the Second World War he had been working with Schaefer on the problem of aircraft becoming charged with static electricity during snowstorms, which could disrupt their radio communications. In the course of experiments on Mount Washington, the home of the 'worst weather in the world' in the northeastern USA, they stumbled across a remarkable phenomenon: whenever a cold wind was blowing, all their equipment became instantly covered with a thin layer of ice. It was clear that under such circumstances the air was filled with tiny droplets of supercooled water that were just waiting for an opportunity to freeze onto a plane's antenna or rigging.

The two researchers promptly abandoned their work on radio communications and devoted themselves to investigating the inner workings of clouds. By that stage it was already common knowledge that water within a cloud didn't simply freeze as soon as the temperature there dropped below zero. But the key question was why? Why, for instance, did some clouds in winter release snow, while other equally cold clouds contained supercooled water droplets that resolutely refused to form ice crystals?

Water droplets in a cloud form around microscopically small 'condensation nuclei' such as specks of dust or soot or salt crystals. The droplets are often so tiny that it takes millions of them to make just one of the raindrops that falls to earth.

If the temperature of the cloud is above freezing point, then a raindrop will be formed by the tiny droplets colliding. However, a cloud will often disperse before the drops have reached the critical size to cause precipitation, and so no rain will fall.

The headline from the *Iowa City Press Citizen* on 14 November 1946 on Schaefer's experiments.

SNOWMAN — Scientist Makes Real Snow in Laboratory; to Try It in Sky from Plane

If the cloud temperature is below zero, the water droplets can freeze into microscopically small ice crystals, onto which other droplets will then freeze in their turn until a flake is formed, which will fall to earth either as snow or, in melted form, as rain. While it is still within the cloud, small ice crystals will detach themselves from the flake, which themselves then become the nuclei for other water droplets to freeze around. Yet this chain reaction evidently doesn't occur in many clouds. Langmuir and Schaefer wanted to find out why.

While Langmuir worked through the problem using theoretical projections, Schaefer attempted to investigate the phenomenon in the laboratory. He lined a deep freeze with black velvet and arranged a spotlight to shine into it in such a way that any ice crystals that formed would be visible from the light they reflected. When he exhaled into the deep freeze, his breath condensed into tiny water droplets at a temperature of –23°C: Schaefer had succeeded, then, in bringing a supercooled cloud into the laboratory.

In a series of over a hundred experiments, he added, variously, volcanic ash, talcum powder, sulphur and other substances to the air in the freezer. Yet, try as he might, he couldn't get any ice crystals to form – until luck lent a helping hand on 13 July 1946. On that morning, Schaefer found that the deep freeze had been switched off by mistake. In order to continue with the experiment as quickly as possible, he put a piece of dry ice into the chest. Dry ice is the solid form of non-toxic carbon dioxide, which freezes at –78°C and at room temperature produces the thick clouds of 'smoke' that are so popular in stage shows.

What had been missing from the cloud of supercooled water droplets in the deep freeze were the initial ice crystals to start the chain reaction that would eventually create snow. If things in the natural world behaved in the same way as inside Schaefer's deep freeze, then these initial ice crystals could easily be produced: all that was required was to cool down parts of the cloud to –39°C. And this is precisely what Schaefer did when he jettisoned his load of dry ice over Pittsfield.

To gain a precise understanding of what happened, exhaustive calculations had to be carried out. To this end, Langmuir engaged the services of the physicist Bernard Vonnegut (brother of the novelist Kurt Vonnegut). His task was to find out how much dry ice was needed to produce what quantity of snow crystals. In making these calculations, Vonnegut hit on a novel idea:

if the initial ice crystals were capable of instigating the chain reaction that led to snow forming, then why shouldn't other substances with a similar form to ice crystals work just as well? Vonnegut studied the crystalline structure of over one thousand substances from tables and finally selected three for testing in the deep freeze. After several failures, one of them – silver iodide – eventually worked. This immediately caused the miniature cloud in the deep freeze to start snowing – but, in contrast to dry ice, at a temperature far higher than –39°C.

So the researchers now had two possible methods of producing the initial ice crystals within a cloud: either ensure that the temperature was below –39°C or sow the cloud with silver iodide crystals.

Meanwhile, Schaefer made a few more test flights with dry ice. One of them was apparently so successful that the legal department at General Electric began to get jittery. At midday on 20 December 1946, Schaefer sowed the clouds above the town of Schenectady in upstate New York with 24 pounds (11 kg) of dry ice. Within two hours or so it began to snow, and it was still snowing eight hours later. This was the heaviest snowfall of the entire winter that year. Although Schaefer was sure he wasn't the cause of 5 inches (12.5 cm) of snow falling, the lawyers at General Electric weren't inclined to take his word for it and banned all tests for the time being.

Try making your own cloud chamber from a plastic bottle and some polystyrene blocks.
madsciencebook.com

Langmuir finally managed to get the US military interested in his research. February 1947 saw the inauguration of Project Cirrus, in which silver iodide was deployed for the first time. This substance had the advantage that it didn't even need to be sown from an aircraft. Researchers could simply produce smoke laced with silver iodide under a promising-looking cloud, and the smoke would then rise up to the cloud of its own accord.

However, the project soon came in for public criticism, and even some of the scientists involved in it accused Langmuir of placing too optimistic an interpretation on his data. For example, in October 1947, he tried to lessen the force of a hurricane by injecting it with a large number of condensation nuclei, so disrupting the storm's momentum. And indeed, no sooner had Langmuir's team dropped their dry ice than the hurricane abruptly made a 90-degree turn. Although such a manoeuvre is nothing untoward for a hurricane, Langmuir was convinced that he was responsible for deflecting it from its original course. Moreover, he later claimed that tests he had conducted in Sorocco, New

Mexico, had caused rain to fall in the Mississippi Delta over 600 miles (1000 km) away, despite the fact that there was no evidence to suggest that the two events were connected in any way.

Langmuir's critics maintained that anyone who knew anything about the weather would regard such a claim as quite fantastic. His adversaries included the US Weather Bureau, an official government body, which had conducted its own experiments in this area and come to the conclusion that cloud seeding was 'of relatively little economic importance'. To this, Langmuir dryly responded: 'The control of a system of cumulus clouds requires knowledge, skill and experience.'

Right up to his death in 1957, Langmuir was adamant that his experiments had worked, yet most scientists remained sceptical, and funding for the project eventually dried up. Although no one was in any doubt that seeding clouds with icy nuclei led to the formation of ice crystals, many experts thought that there was no basis for claiming that this ultimately caused more rain to fall. Langmuir's statistical analysis was inadequate, and right up to the present day, data-processing remains a real bugbear for rainmakers, since (unlike in Schaefer's deep freeze) you can never be sure when conducting experiments in the atmosphere whether the rain might not have fallen in any case, regardless of any cloud seeding.

After Project Cirrus came to an end in 1953, Schaefer went on to work on various meteorological problems. He died in 1993 at the age of 87 in Schenectady – the same place where he had caused – or not, as the case may be – the heaviest snowfall of the winter almost half a century earlier.

Research is still going on today, albeit on a smaller scale, into creating the weather to order. For instance, the Weather Modification Association holds an annual conference, and there are a number of research teams who are intent on doing experiments on a firm statistical basis. Yet all those working in this field are now clear on one thing: the weather is too complicated to be susceptible to manipulation with crude methods.

This was recently brought home forcibly to the Russian president Vladimir Putin, who wanted to ensure fine weather for the celebrations marking the 300th anniversary of the founding of St Petersburg. A budget far in excess of half a million euros had been set aside for ten aircraft of the Russian Air Force to seed any approaching rain clouds prior to the festival. As the Russian

Source

Schaefer, V.J. (1946), 'The Production of Ice Crystals in a Cloud of Supercooled Water Droplets'. *Science* 104 (2707), pp. 457–459.

Meteorological Service announced, the pilots' mission was to 'prevent the rain from ruining the ceremony on the Neva'. But on the day, as Putin prepared to welcome his official guests in front of the statue of St Peter and conduct them on a walkabout to St Isaac's Cathedral, the heavens opened.

Such fiascos no doubt explain why there are now so few commercial enterprises offering such services as hail prevention, rainfall increase or the dispelling of fog. Manipulating the weather is a tough line of work to be in, since a success can cause as many problems as a failure: for example, when the US brewer Coors had the clouds above its barley fields seeded in 1978 to guard against the possibility of hail, other farmers in the area began a court action, since they suspected that Coors's real agenda was to prevent any rain from falling during the harvest period. The judge found in their favour, and Coors was forced to call off the operation.

1946 Holidaying in a Draught

Anyone wanting a cheap holiday in Great Britain just after the Second World War made a beeline for Salisbury. At that time you could find free accommodation just outside this small Wiltshire cathedral city in spacious flats, each housing two people, and fully equipped with books, games, a radio and telephone. You could spend your leisure time playing table tennis, badminton or golf; and you even got paid three shillings a day for your trouble.

There was just one small hitch: the buildings of the Harvard Hospital high up on windy Salisbury Plain a few miles outside the town were home to the British government's Common Cold Unit, and the visitors – students, in the main – acted as guinea pigs. Although they were 'unsatisfactory' – as the head of the Common Cold Unit, Christopher Howard Andrewes, put it in an article in 1949 – they were 'the only animals available'.

At that period, the only other animal apart from man that could be infected with cold germs was the chimpanzee. But according to Andrewes, chimps were 'very expensive, very strong and difficult to handle'. Not so the students, and the 'ten-day free holiday' in the Harvard Hospital was a popular institution. Many of the test subjects came back several times.

The 12 participants who were required to spend half an hour in a draughty corridor after taking a hot bath one Saturday morning presumably weren't among those who returned for more. They felt 'chilly and miserable', Andrewes wrote, and their mood probably wasn't improved by the wet socks that they were made to wear for the rest of the morning.

Protective suits like this were used at the Common Cold Unit in Salisbury to guard against infection.

If one subscribed to the widely held view, then this sort of treatment was a sure-fire recipe for catching a cold. And it was precisely such popular attitudes that Andrewes was determined to subject to scientific scrutiny, since there were certain observations that flatly contradicted them. For example, Arctic explorers who went on long expeditions never caught a cold. Moreover, the people who lived in Eskimo villages never fell ill in winter, when the weather was at its coldest; it was in the spring that they became ill, after the first foreign ships of the year had visited.

The test subjects in the wet socks had arrived in Salisbury three days earlier. Like all other experiments conducted at the Common Cold Unit, the one they were involved in began on a Wednesday. The 18 test subjects underwent an initial medical examination before taking up residence in the flats set aside for the experiment, two people to each apartment. They were instructed to maintain a distance of at least 30 feet (9 m) from all unprotected persons other than their flatmates for the next ten days. They were allowed to take a walk, but told to avoid entering buildings or getting into cars. When examining them, all doctors and nurses wore protective clothing and face masks. Their food was left in insulated containers outside their apartment doors three times a day.

The period between Wednesday and Saturday passed pretty uneventfully. The purpose of this grace period was to distinguish any cold that a test subject might have brought with him or her to the unit from the actual experiment.

Then, on Saturday morning, the doctors divided the participants up into three groups, each comprising six people. The first six had a filtered and diluted nasal secretion from a cold sufferer dripped into their noses. The second six were subjected to the chilling treatment using the bath, the draught and the wet socks. Finally, the third group of six had both the chilling treatment and the nasal secretion administered simultaneously.

Scientists at the time were pretty sure that colds were caused by viruses, which were found primarily in nasal secretions. Since the main symptom was a very runny nose, the researchers at Salisbury took the daily increase in the weight of tissues a person used as an indicator of how heavy a cold they had. However, in order to be able to determine precisely what the pathogen was, this would have to be cultivated in the lab, a task that turned out to be extremely difficult.

A few days after their treatment on the Saturday, the first participants came down with a cold: four of the test subjects who had been given both the chilling treatment and the virus, plus two who had only been infected with the virus. Chilling on its own gave no one a cold.

This seemed to bear out the popular wisdom: although coldness alone couldn't give you a cold, it clearly encouraged

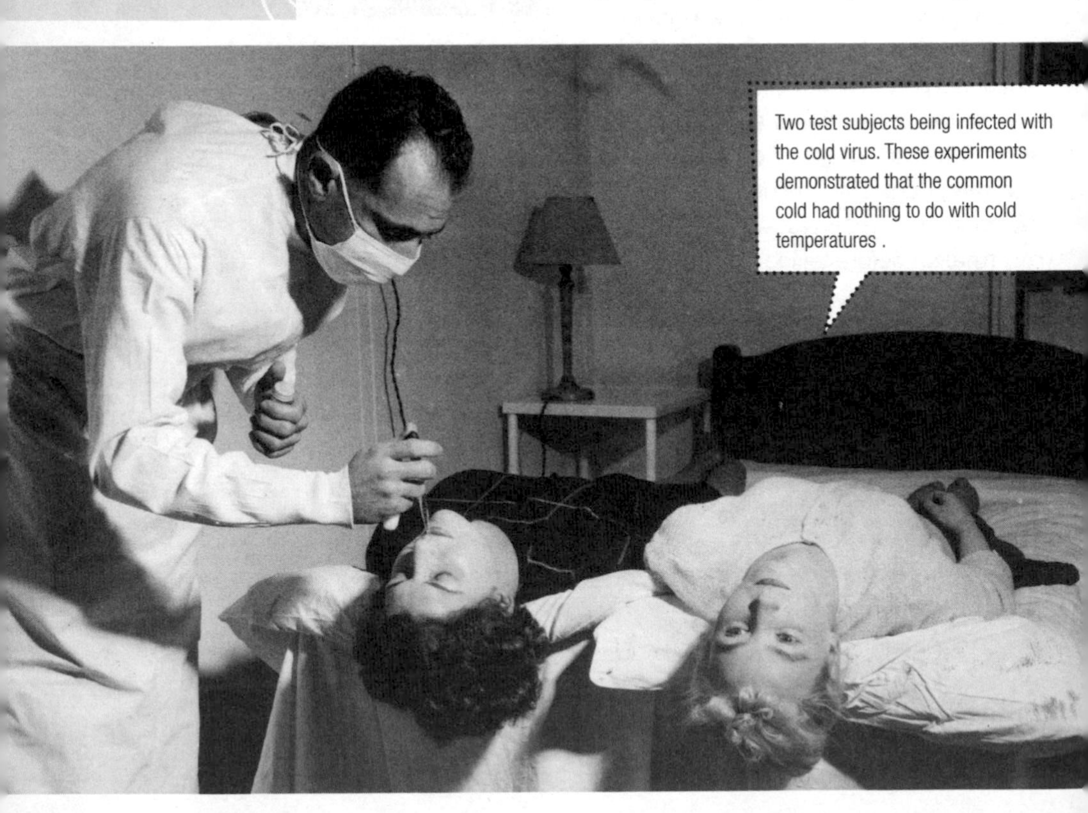

Two test subjects being infected with the cold virus. These experiments demonstrated that the common cold had nothing to do with cold temperatures .

the spread of the virus. Yet Andrewes wasn't content to leave it at that, since the number of participants involved in the tests was simply too small to enable him to draw any hard-and-fast conclusions. As he explained, 'We were foolish enough to repeat this experiment, with a contrary result.'

Once again, the results showed that chilling alone produced no cold symptoms, but this time twice as many people from the group that had only been infected with the virus came down with a cold compared to those who had also been made to stand out in the cold. A third go at the experiment also produced the same outcome: once again, there appeared to be no link between catching a cold and having been exposed to a cold draught beforehand.

Andrewes's experiments were the first in a series of such tests that took place in the 1950s and 1960s, involving hundreds of participants. Not one of these experiments was able to demonstrate that chilling a person down had any connection whatever with their contracting a cold.

To this day no one has fully explained why colds occur more frequently in our latitudes in winter than they do in summer. Further studies have proved that this greater frequency clearly has nothing to do with a weakened immune system or the dry air in heated rooms. Rather, experts believe that the virus spreads more easily when people are in close proximity to one another in badly ventilated rooms in winter. In addition, sunlight – the ultraviolet component of which kills germs – is not so strong in the winter.

Yet even Andrewes in his day was well aware that science faces an uphill struggle against well-established folk wisdom: 'Even the most eminent men of science,' he wrote, 'almost invariably lose all sense of critical judgment when colds and especially their own colds are concerned.'

All the more so when the cause of a particular illness seemed to be indicated in its very name.

madsciencebook.com Amusing historical footage of the experiment.

Source

Andrewes, C.H. (1948), 'Cantor Lecture: The Common Cold'. *Journal of the Royal Society of Arts* 103, pp. 200–210.

1948 Spiders 1: Drug-induced Webs

Spiders have a habit that makes it very arduous for researchers studying them: they invariably spin their webs at around 4 o'clock in the morning.

Peter N. Witt administering drugs to a spider.

One of those who found himself grappling with this problem was the zoologist Hans M. Peters from Tübingen University in Germany. In 1948 Peters planned to film the spiders building their webs but wanted to avoid having to get up in the middle of the night every time he did so. And so he approached Peter N. Witt, a young research assistant in the university's pharmacology department, and asked him whether it would be possible to administer stimulants to spiders to induce them to construct their webs at a less ungodly hour. Witt first tried giving them strychnine, morphine and dextroamphetamine ('speed'). The process of administering these to the spiders was quite straightforward: so long as it was mixed with a small amount of sugar solution, the spiders would ingest any toxin you cared to give them. But all to no avail. The spiders continued to work at the crack of dawn, and Peters lost all interest in the experiments.

Witt, on the other hand, was fascinated by the results: the webs that the spiders spun while under the influence of drugs were such as he had never seen before. There were extremely gappy ones, closely knit ones, grotesquely irregular ones, yet also some that were made extremely meticulously. Could spiders' webs, then, be used as a way of measuring the effects of drugs and other medication? At that time hardly any procedures existed for quantifying the effect of such substances on an organism.

And so Witt proceeded to dose the spiders with everything he could lay his hands on in the medicine cabinet: mescaline, LSD, caffeine, psilocybin ('magic mushrooms'), luminal and valium. Then he let them build their webs in a square frame measuring 14 x 14 inches (35 x 35 cm), and photographed them against a black background.

Because the webs, when seen with the naked eye, didn't fit into any obvious categories, Witt devised a statistical method that was capable of distinguishing between even the slightest systematic variations in their construction. By studying the photographs of

Source

Witt, P.N. (1956), *Die Wirkung von Substanzen auf den Netzbau der Spinne als biologischer Test* ('The effects of various substances on spiders' web-building as a biological test'). Springer.

the webs, he determined such factors as angles, distances between individual threads and the area covered, and then drew up tables indicating the frequency with which webs were made, the size of their capture surfaces and the ratio of the webs' axes to one another.

Unsuitable results for use in drug prevention: the most chaotic web (left) was made under the influence of caffeine, while the most beautiful (right) was produced by a spider on marijuana.

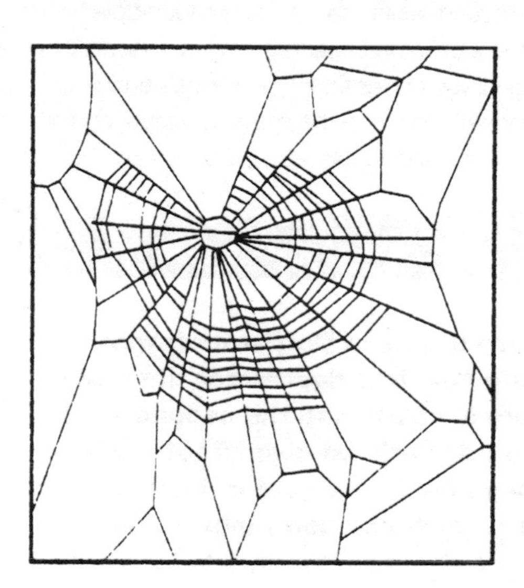

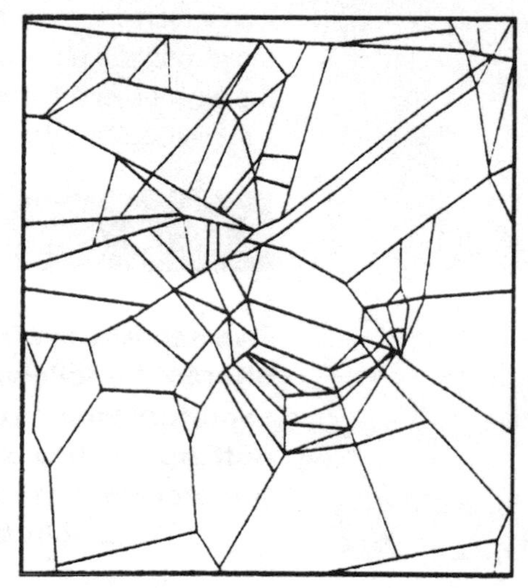

This was an extremely laborious task: the web of a fully grown female Araneus diadematus could easily comprise as many as 35 radial threads and 40 capture spirals. Such a web would have no fewer than 1400 junctions. In order to draw any meaningful comparison, 20 webs needed to be analysed before administering the drug and 20 after. At that time, before the advent of computers, it was nigh-on impossible to process such a huge amount of data. To make things easier for himself, Witt therefore confined himself to measuring only those areas that looked especially interesting after a particular drug had been administered. However, this in turn made comparison between the effects of the various substances more difficult.

After further bizarre experiments of this kind (see p. 139), scientists abandoned the idea of trying to use spiders' webs as a universal indicator of the effects of chemical substances. Later studies were no longer concerned with identifying the drug that had been administered from the configuration of the web, but rather with the effect of a particular drug on the spider's nervous system. In 1995, scientists from NASA published the results of their tests (why NASA, of all people, should carry out such experiments is anyone's guess). By this stage, computer technology had come on in leaps and bounds, and the webs could be analysed using statistical programs that had been specifically designed for crystallography. One thing emerged very clearly from these results: what the spiders produced was worse than useless for any drug prevention programme. The most chaotic webs were spun under the influence of caffeine, the most beautiful with marijuana, and the most perfectly regular – something that Witt himself had found – by spiders on LSD.

1949 The Secretaries' Deal

Two secretaries from the Rand Corporation were offered the following deal one day: either the first secretary would receive $100 and the second nothing, or they would jointly be given $150 – with the proviso that they could agree beforehand on how to split this sum between them. This game was the brainchild of the mathematician Merrill Flood. Flood wanted to find out how people would divide up a win if they had the chance of getting extra money by cooperating.

Source

Poundstone, W. (1992), *Priisoner's Dilemma*, Doubleday, p. 102

He predicted that the first secretary would take $125 dollars and the second $25. In this case, both would be $25 dollars better off than if they hadn't cooperated. Otherwise, the first secretary would have received $100 and the second nothing. And yet the secretaries took a different line. They split the total sum down the middle, so that each of them got $75. Evidently, Flood concluded, people don't act solely according to the mathematical logic of maximizing their profit. Rather it was the case, in situations such as the one Flood had devised, that social relationships had a strong bearing on their behaviour.

1949 Orgasms in a Staccato Rhythm

When the medical researchers Gerhard Klumbies and Hellmuth Kleinsorge from the Jena University

Clinic published the results of their experiments, out of a sense of decency and propriety they formulated certain passages in Latin. It was common practice in medical literature in those days not to describe certain intimate details in German.

In their article they claimed that a 'strange circumstance' had made their investigation possible. Exactly what they meant by this was intelligible only to those with a good enough classical education to understand what a 'Femina supersexualis, quae emotione animae se usque ad orgasmum irritavit' is: namely, a woman who can bring herself to orgasm just by fantasizing. The woman in question, who was in her thirties, had attended the clinic because she was concerned about her ability to bring herself to a climax any time and any place just by squeezing her thighs together. Klumbies and Kleinsorge quickly realized that this ability gave them a unique opportunity to find out more about the strain placed on the body during orgasm. They reassured the woman and asked her permission to measure her pulse and blood pressure. She agreed.

It was already known at that time that sexual intercourse could induce strokes and heart attacks, which in some cases were fatal. And yet it remained 'unclear what stresses the whole organism undergoes during the act of coupling', as the doctors wrote in their paper, going on to state that 'anyone who imagines that they can estimate the amount of strain merely by observing external evidence is very much mistaken'. Accordingly, they did not propose estimating the stresses involved, but rather intended to measure them.

The 'Femina supersexualis' was the ideal test subject. She could produce orgasms to order and in quick succession, and simply lay there calmly all the while she did so. No shaking disturbed the sensitive measuring instruments, and no sudden movements caused the woman to get tangled up in the mass of cables that led off into an adjoining room. There, Klumbies and Kleinsorge sat intently watching the pen on the recording device and monitoring the changes in her blood pressure.

The first orgasm cited in their paper caused the woman's systolic blood pressure to increase by 50 to a level of 160 mmHg. This was a 'remarkable' rise, given that this figure was one-fifth higher than that recorded in women who were suffering labour pains. To give themselves some basis for comparison, they made the

Source

Klumbies, G., and Kleinsorge, H. (1950), 'Das Herz im Orgasmus' ('The Heart during Orgasm'), *Medizinische Klinik* 45, pp. 952–958.

woman – 'a trained athlete' – run up the six-storey staircase in the clinic, which only caused her blood pressure to rise by 25 mmHg. Her pulse increased to 98.

In the course of another test, the woman produced sequences of orgasms at one-minute intervals. All the while, her pulse and blood pressure readings showed a similar progression. In the first 5 seconds, her pulse rate rose sharply by around ten extra beats per minute, then remained at this level for the next 15 seconds before increasing during her orgasm, which took place after about 25 seconds, by a further five beats per minute. In the course of these experiments, her blood pressure rose to over 200 mmHg.

In order to compare the physical reactions of men and women, Kleinsorge and Klumbies also recorded the same data for a man during orgasm. Yet the patient in question, who had come into the clinic to have his fertility investigated, wasn't blessed with the same remarkable aptitude as the woman. He took 15 minutes to bring himself to orgasm by masturbating, and in his case Kleinsorge and Klumbies had to content themselves with a single measurement. The figures that were recorded showed that the male orgasm is considerably more stressful than that experienced by women. His pulse increased to 142 and his blood pressure to 300. It was presumably for this reason, speculated the authors, that 'the rumour persists that the female orgasm is a myth'.

However, the similarity in the pulse and blood pressure graphs for men and women might not just be mere coincidence. For men and women alike, the orgasm represents the 'peak in blood pressure, heartbeat volume and pulse rate'. The steady curves appearing on the graphs, which were then interrupted by a sudden peaking surge at the moment of orgasm, encapsulated for the two doctors the old adage that 'the pleasure all lies in the getting there and not in the arriving'. They continued: 'We have been able to ascertain that where sexual pleasure is concerned, the former counts for everything and the latter for nothing.'

The paper's authors didn't seriously believe that their work would lead to any effective measures being taken to prevent 'unfortunate incidents during the sex act'. As they put it, 'It's a futile undertaking to try to ban sex. The doctor may well be stronger than Bacchus, but there's no question he's weaker than Venus.'

" unfortunate incidents
during the sex act "

1950 Be Good-natured – but not a Sap!

One afternoon in January 1950, the two mathematicians Merrill Flood and Melvin Dresher invited two of their colleagues to take part in a game they had invented that very morning. The game wasn't particularly intellectually challenging – you just needed to respond to questions with the answer A or B. None of the participants that day could have guessed that politicians and generals would soon start to take an interest in what they were up to. In each round, each of the two players secretly plumped for either A ('cooperate') or B ('don't cooperate'). Once both had made their choice, they divulged what it was. Depending on what they had both selected, the players were either rewarded or fined. The idea was to play a hundred rounds of the game.

The field of study that is concerned with games of this kind was still in its infancy at that time: it goes by the name of 'game theory', and is a mathematical technique for analysing conflicts. For instance, in a business context, say, a buyer's prime motivation is to pay as little as possible, whereas a seller wants to charge as much as he can. And yet the final price paid isn't strictly governed by the laws of supply and demand. The decisions that people come to can sometimes be quite irrational and not aimed at maximizing their profit. A typical scenario for conflicts as viewed through game theory is this: at the moment when a decision is taken, none of the competitors knows what the others will do, and yet they are all fully aware that the final outcome will be determined by the decisions that each and every one of them arrive at.

Indeed, this was also true of Flood and Dresher's game. In each round the two players, Armen Alchian and John D. Williams, were faced with four possible outcomes, namely: both players chose to cooperate; both players chose not to cooperate; Alchian chose to cooperate but Williams didn't; or vice versa. They could consult a table to see which of them would get how many cents in each instance. And as they looked at this table, it soon dawned on them wherein the real dilemma of this game lay: if both of them cooperated, then Alchian got half a cent and Williams a cent. If neither cooperated, each of them received half a cent less, meaning Alchian got nothing and Williams half a cent. Considered in this light, cooperation appeared to be the best strategy for both of them. But the real problem arose with the other two options, since the person who chose to cooperate in isolation was penalized, whereas the one who chose not to cooperate was rewarded.

So, if Alchian cooperated and Williams didn't, Alchian had to pay a fine of a cent, whereas Williams was given two. In the opposite instance, Williams had to pay a cent, and Alchian got a cent. In these cases, the confusing tariff of different rewards and penalties had no bearing on the basic dilemma facing the players.

Since neither player knew what tactics his opponent would adopt, he could only conclude that non-cooperation was the only sensible course of action: in these circumstances, then, the best-case scenario would be that the other player chose to cooperate and so you reaped the reward. Likewise, in the worst case, the other player would choose not to cooperate as well, meaning that at least you didn't lose out. This was precisely the modus operandi that the mathematician John Nash, in his minimax theory, predicted two rational players would adopt.

Source

Flood, M.M. (1952), 'Some Experimental Games' (RM-789). RAND Corporation. In: Poundstone, W. (1992), *Prisoner's Dilemma*. Doubleday, p. 108.

And yet this apparently rational behaviour actually leads to a paradox: ultimately, both players should logically come to the same conclusion and never cooperate. But at some stage it would dawn on them that this was generating less income for them than if they both acted 'irrationally' and cooperated all the time. Evidently, logic ran counter to the best outcome for all concerned.

As Flood and Dresher anticipated, Alchian and Williams didn't stick to Nash's theory, but instead began to act irrationally, with Alchian cooperating in 68 out of 100 rounds, and Williams in 78.

Flood and Dresher published the result of their strange experiment in an internal research memo. Anyone reading it carefully could have foreseen even at this stage the brilliant future that lay ahead for this experiment. Indeed, this is also evident in the notes that Alchian and Williams jotted down after every round: 'Probably learned by now.' 'The stinker!' 'He's crazy. I'll teach him the hard way.' 'Let him suffer.' 'Maybe he'll be a good boy by now.' 'To hell with him.' 'He requires great virtue, but doesn't have it himself.' 'Goodness me! Friendly!' 'This is like toilet training a child. You have to be very patient.'

The game turned on trust and betrayal. Some commentators later identified in the dilemma it posed the fundamental problem facing society: individuals or groups whose actions benefit themselves but have a highly detrimental effect on the common good.

However, it wasn't the complicated version with the confusing rewards in cents that made the experiment popular. A colleague of Flood and Dresher's, Albert Tucker, repackaged the dilemma within a different narrative and gave it the name that was to make it famous: the prisoner's dilemma. One version of it runs as follows. Two members of a gang are arrested and interrogated separately. But the police lack the evidence to charge both of them with the main crime. A more minor crime that can be proved without any further evidence would see both men sent down for a year. And so the police offer each of the accused a deal: if one of them is prepared to testify against the other, he can go free, while his accomplice will go to prison for three years. But the catch in this is that if they both testify against each other, then they'll both go down for two years apiece.

A rational suspect would think the situation through in the following way: if I shop my mate and he keeps his mouth shut, then I'm out straight away rather than having to spend a year in prison (which is what I'd get if I kept quiet). If we both squeal on each other, I'll go down for two rather than three years (which is what I'd get if I kept stumm and my mate shopped me). So, in any event, I'm better off if I spill the beans. The only problem is that the other suspect will surely come to the same conclusion and so both of them will find themselves locked up for two years. If only both of them had kept quiet, then they'd only be in for one year.

The prisoner's dilemma always comprises the same ingredients: a reward if all those involved cooperate, a penalty if nobody cooperates, and a temptation that one might potentially maximize one's gain if one doesn't cooperate and everyone else does.

The whole world is made up of prisoner's dilemmas. Shoplifting, tax evasion, fare dodging – as long as all the others pay, you can get away with it, but if everyone decides not to pay, then you'll all be punished.

The example par excellence of the prisoner's dilemma was the arms race between the USA and the Soviet Union. Admittedly, Flood and Dresher didn't have this in mind when they devised their game, but the parallels are obvious – after all, both mathematicians were working for the Rand Corporation, a research institute in Santa Monica near Los Angeles that had close links to the US military.

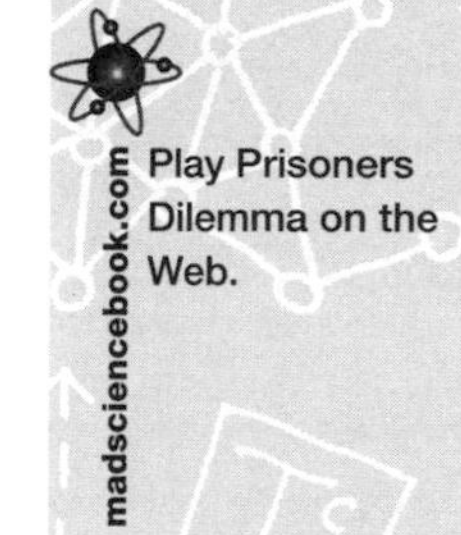

When two nations are in the process of deciding whether to build up an arsenal of nuclear weapons, their reasoning runs thus: if the other power is the only one making nuclear weapons, then we'll be at a disadvantage. So they both construct nuclear weapons. But if both act in this way, then the advantage they hoped to gain actually becomes a disadvantage. Things would have been more secure if neither side had developed them in the first place.

Since Flood and Dresher's original experiment in 1950, the prisoner's dilemma has had a remarkable career. Hundreds of pieces of research in mathematics, business studies, psychology and biology have been based around it. In the strictest sense, it doesn't have a solution – otherwise it wouldn't be a dilemma – yet game theory has given us the ability to describe conflicts in detail and devise strategies for dealing with them.

And so non-cooperation is only the best strategy when adversaries meet only once. In cases where they encounter one another repeatedly – for instance, people who do business with one another on a regular basis, or apes who delouse one another – it's wise to take a different approach.

In 1979 the political scientist Robert Axelrod tried to find out exactly how one should best proceed in such cases. He had game theorists confront their colleagues with their various preferred strategies. To everyone's surprise, the simplest strategy came out on top, namely, cooperate in the first round, then do precisely what the other player did in the previous round. This strategy was dubbed 'tit-for-tat'.

This strategy was refined in subsequent experiments. According to the findings, a somewhat more altruistic approach appeared to be more successful than 'tit-for-tat'. In other words: if someone cheats you, hit back immediately but then forgive your adversary and try cooperating once more. Even more plainly put: be good-natured – but not a sap.

1951 Nosediving in the Vomit Comet

As ever-higher-performance jet aircraft were developed at the end of the 1940s, so the need became more pressing for aviation medicine to simulate the various phases of flight.

American astronauts during a parabolic flight in 1959. The zero-G flight path in a jet remains to this day the only method of experiencing half a minute of weightlessness within Earth's gravitational field.

Centrifuges could recreate the stresses placed on the human body by rapid acceleration, while pressurized chambers mimicked the decrease in air pressure at high altitudes. Scientists also assumed that weightlessness might well present serious problems for the human body. 'However, confronted with the necessity of producing states of sub-gravity, we must concede that all tricks of simulation fail,' wrote Fritz and Heinz Haber from the USAF School of Aviation Medicine at Randolph Air Force Base in Texas. In their legendary article 'Possible Methods of Producing the Gravity-Free State for Medical Research' they ran through all possible scenarios and finally ended up devising a method after all – albeit not one that could be put into practice on the ground.

At this stage, the paper's authors didn't have the future astronauts of the US space programme in mind. The era of space travel was still too far off, and although jets were reaching extremely high altitudes, gravity there was still only marginally lower than at ground level. Rather, what prompted their deliberations was the fact that certain flight manoeuvres produce a momentary state of weightlessness. For instance, if an aircraft's engines suddenly cut out at high altitude, it would plummet earthwards in free fall and the pilot would become weightless.

Source

Haber, F., and Haber, H. (1950), 'Possible Methods of Producing the Gravity-Free State for Medical Research'. *Journal of Aviation Medicine* 21, pp. 395–400.

To this day, nobody has managed to create a machine that could produce even the slightest reduction in gravity at ground level, let alone eliminate it entirely. Despite the fact that some researchers continue to claim that they have achieved this feat, most physicists regard it as impossible. Under the influence of the Earth's gravitational pull, gravity can only be overcome with the aid of movement, claimed Fritz and Heinz Haber – say in a falling lift. Assuming that the lift isn't slowed down by air resistance, it will fall at exactly the same rate as the person inside it, who is weightless for the duration of the fall. The problem with this was that, even in an extremely tall lift shaft, the fall wouldn't last very long. Yet the concept of the falling lift helped put the authors on the right track: irrespective of how a person falls through the air, if he is confined within a cabin that is moving in exactly the same direction as his flight path, he will be weightless inside it. In order to ensure that this phase lasted as long as possible and that the fall could be cushioned, the Habers concluded that this cabin would have to be an aircraft following a wave-like flight path. The plane would initially climb at an angle of 45 degrees and then slowly decrease its angle of ascent until it reached the highest point of the climb, after which it would follow a symmetrical path downwards. This was exactly the parabola that a person would describe if he were to be shot into a vacuum by a catapult set at a 45-degree angle. The authors predicted that this experiment would produce up to 35 seconds of weightlessness.

In the summer and autumn of 1951, the test pilots Scott Crossfield and Charles E. ('Chuck') Yeager discovered that the Habers were quite correct in their assumption: by flying their jet fighters in a parabola, they attained a state of weightlessness for up to 20 seconds. Afterwards, Crossfield reported that zero-G had made him light-headed but had not affected his powers of coordination. Yeager had the impression of being in free-fall and felt 'lost in space'.

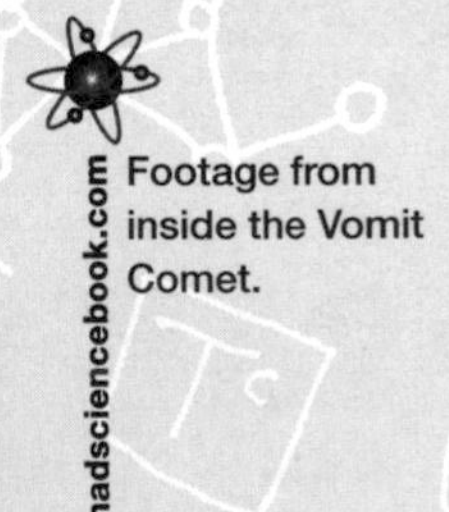

Footage from inside the Vomit Comet.

Parabolic flights have remained to this day an integral part of astronaut training. However, would-be astronauts aren't strapped into jet fighters like Crossfield and Yeager, but are instead taken up in a spacious, specially equipped KC-135 aircraft belonging to NASA, which has a padded cabin. Thanks to the typical physical reactions experienced by its occupants, this plane has been given the nickname 'the Vomit Comet'.

Anyone who doubts that people in the cabin of the KC-135 really attain a state of weightlessness during parabolic flight should take a look at the film *Apollo 13* starring Tom Hanks. The film crew hired out the actual 'Vomit Comet' to shoot the relevant scenes.

1951 Twenty Dollars for Doing Nothing

The appeal that went out sounded like easy money: the psychologist Donald O. Hebb from McGill University in Montreal was on the lookout for students who were willing to do nothing for $20 a day. All they were required to do was to lie on a bed in a soundproofed, brightly lit room, and to wear mitts on their hands, cardboard tubes on their forearms and glasses that allowed only diffuse light to filter through them. They were allowed to get up for meal times and toilet breaks only, but not to take the glasses off for either of these functions.

For a long time, Hebb had been mulling over the question of what happened to the brain when it was cut off from all the stimuli it would usually receive. The prevailing theory was that, in order to function normally, it needed a variety of sense impressions. In animals, the kind of total isolation Hebb was after could be achieved simply by severing the brainstem. 'College students, however, are reluctant to undergo brain operations for experimental purposes, and so we shall have to content ourselves with a less extreme isolation from their environment,' Hebb remarked in his notes on the experiment. These also cited a practical reason for conducting the experiment: people who have monotonous jobs, such as watching radar screens, tend to make mistakes, and researchers were keen to find out what the precise causes of such mistakes might be.

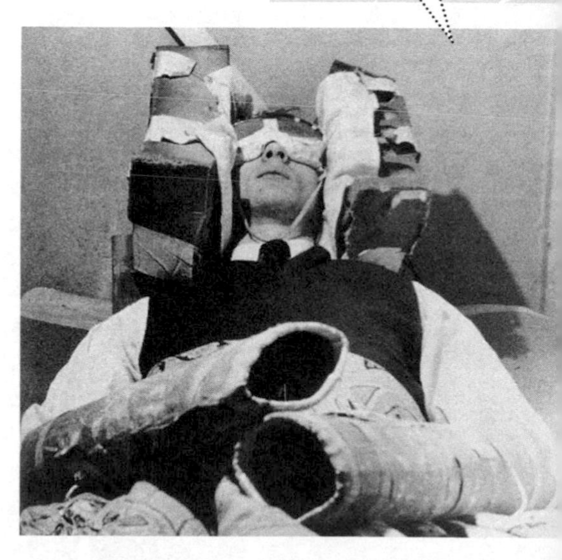

What happens when the brain is screened off from all external stimuli? This participant in an isolation experiment at McGill University in Montreal experienced hallucinations.

However, the true catalyst of the experiment wasn't mentioned. This was the fact that the Soviet Union and China were using sensory deprivation techniques in their brainwashing of prisoners. Consequently, the military were extremely interested in Hebb's experiments.

Twenty-two participants were quickly found. Yet none of them stuck it out in the test room for longer than three days. Although their $20 fee was more than double what they could otherwise hope to earn in the same period, the psychologists had their work cut out persuading them to stay. In fact, the test subjects had planned during their isolation to learn facts by rote, to prepare seminar papers or to map out a forthcoming lecture, but in practice they all reported that after a certain amount of time they were no longer able to think about a particular topic with any degree of concentration. As one participant put it, 'I just run out of things to think of.' Some began to count aloud out of sheer boredom.

The students finally abandoned themselves to daydreaming, and let their thoughts wander. Psychological tests indicated that isolation seriously impaired the capacity to think. But the most important result was an unexpected side effect: all the test subjects experienced hallucinations. All of a sudden, they saw changes in colour and wallpaper-like patterns, but also complex scenes, such as prehistoric animals in a jungle setting or a procession of squirrels carrying sacks over their shoulders and trudging through a snowy landscape.

Hebb's isolation experiment established a new field of research. In the years that followed, hundreds of similar experiments were conducted. Not only the military but also NASA was interested in the results, since it reckoned that similar situations such as those in Hebb's test room might well arise during long space flights.

Four years after the first experiments in Montreal, an eccentric researcher in the USA hit on a novel way of heightening the sense of isolation from the outside world, and in the process dramatically enhancing the participants' hallucinations (see p. 140).

Source

Bexton, W.H., et al. (1954), 'Effects of Decreased Variation in the Sensory Environment'. *Canadian Journal of Psychology* 8, pp. 70–76.

1952 Spiders 2: Spinning a Web with Amputated Limbs

Anyone who's still troubled by their conscience because they pulled the leg off a spider when they were a kid can take solace from the 48-page study

published by Margrit Jacobi-Kleemann. This biologist 'amputated' different numbers of legs from several spiders of the species Araneus diadema before filming their efforts to spin webs with a cine camera. After studying around 10,000 separate frames, she came to the conclusion that 'after the loss of one or more legs, Araneus diadema is still capable of constructing a fully functioning web for trapping insects'. Yet Jacobi-Kleemann had cut off only two legs at most: one on the left side and one on the right. Anyone who went further than this in their childhood and committed more monstrous acts won't find any absolution from the world of science.

Source

Jacobi-Kleemann, M. (1953), 'Über die Lokomotion der Kreuzspinne Aranea diadema beim Netzbau'. ('On locomotion in the garden spider Aranaeus diadema during web building'). *Zeitschrift für vergleichende Physiologie* 34, pp. 606–654.

1954 Frankenstein for Dogs

Most visitors to the State Museum of Biology in Moscow pass by the glass case devoted to transplant medicine without giving it a second look. At first glance, it's hard to make out quite what a monstrosity is being displayed within – it just appears to be a stuffed puppy lying directly in front of a fully grown dog, as if there hadn't been enough room to display them properly next to one another.

In actual fact, the puppy's body stops immediately behind its front paws. It was at that point that the Russian surgeon Vladimir Demikhov sewed the animal's truncated body onto the adult dog's neck.

Demikhov presented his work to the Moscow Surgical Association on 26 February 1954. Eight years previously, he had carried out a heart transplant operation on dogs, and had subsequently conducted lung transplants and bypass operations as well. The operation with the dogs' heads was now the world's first transplant that involved an entire system comprising different organs, Demikhov claimed. In a three-hour procedure, he cut the puppy's body in two between

Even local papers like the *Lethbridge Herald* (16 December 1954) ran articles on the two-headed dog.

Scientist Claims He Can Produce 2-Headed Dogs

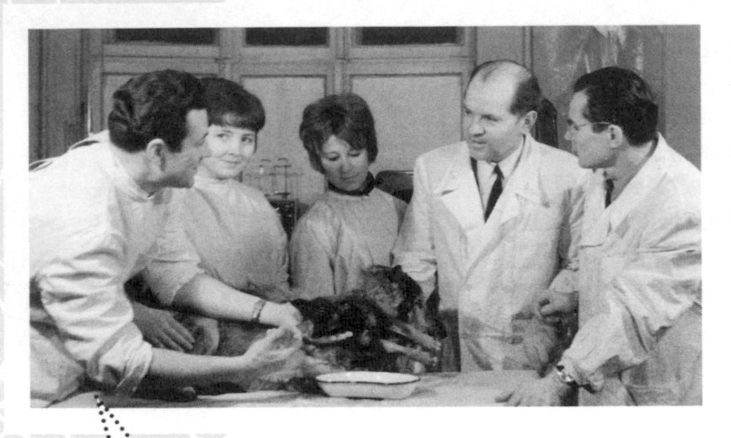

The surgeon Vladimir Demikhov (second from right) with his bizarre creation – a four-year-old mongrel with the head and forepaws of a two-month-old puppy grafted onto its body.

the fifth and sixth ribs – leaving behind its heart and lungs – before linking its arteries and veins to those of an Alsatian, and finally affixing the puppy's head to the Alsatian's skeleton. He left its windpipe and its oesophagus open: the blood supply to the puppy was maintained by the Alsatian's circulatory system. Three hours after the operation, the Alsatian blinked, and after a further four hours, it moved its head. After a day had passed, the transplanted puppy's head had also

One of Demikhov's two-headed dogs is still on show at Moscow's State Museum of Biology.

regained its strength, biting one of Demikhov's assistants so hard in the finger that it drew blood.

This pitiful monstrosity died of an infection after six days. But Demikhov didn't let this setback deter him. Over the following years, he carried out 20 more such operations. In one instance, he even transplanted a young dog onto its mother's body. The record survival time for one of his creations was 29 days, in 1959.

Even at the time, there was controversy over what insights were supposed to be gained from these experiments, but they did succeed in raising Demikhov's profile around the world. After the Soviet Union's launch of Sputnik I in 1957 made it the first nation to put a satellite into orbit, Demikhov's operations were hailed as the 'Sputnik of surgery'.

Source

Demikhov, V.P. (1962), *Experimental Transplantation of Vital Organs*. Consultants Bureau, pp. 162–170.

1955 Spiders 3: Urine in the Web

In 1948 the pharmaceutical researcher Peter N. Witt discovered quite by chance that spiders build quite different webs when under the influence of drugs than they do otherwise (see p. 123). The psychiatrists at the Friedmatt Sanatorium and Nursing Home in Basle, Switzerland, were aware of Witt's work and hit on the idea of trying to get to the bottom of schizophrenia – using spiders.

It was a mystery – and remains so to this day – what the precise trigger was for the onset of this mental illness. However, 50 years ago scientists thought that they had found a promising lead: after taking drugs such as mescaline or LSD, healthy patients began to show symptoms similar to those exhibited by schizophrenics. These chemical substances induced short-term hallucinations and personality disorders. Could it be that such substances were permanently present in the metabolism of those suffering from schizophrenia? In other words, were schizophrenics on a constant 'high' due to a mere whim of their body chemistry?

So, at the start of the 1950s, researchers in Basle began to examine the urine of schizophrenics in an effort to discover what this chemical compound might be. Urine was chosen as the basic material for their investigations 'so that we'd never be stuck for large quantities to work on', as one of the team involved later wrote. But how on earth were they supposed to find a substance

Source

Rieder, H.P. (1957), 'Biologische Toxizitätsbestimmung pathologischer Körperflüssigkeiten' ('Biological determination of toxicity in pathological bodily fluids'). *Psychiatria et neurologia* 134, pp. 378–396.

without knowing what it consisted of, and which they weren't even sure existed in the first place?

The biologist Hans Peter Rieder collected and prepared 50 litres' worth of urine samples from 15 schizophrenics. The resulting urine concentrate was fed to spiders, and the webs that they spun were then compared to those constructed by spiders that had been given researchers' urine instead. If any systematic difference was evident in the webs made by these two groups, then it might well be that the substance they were trying to find was responsible. Moreover, if the webs also resembled those spun by spiders under the influence of LSD and mescaline, then the scientists would at least know what type of substance they were looking for.

The experiment was conducted several times with various different concentrations of urine, but the results were disappointing: although the spiders certainly constructed different webs when under the influence of urine than they did otherwise, no systematic difference was apparent between researchers' and schizophrenics' urine. After a further series of experiments, the team came to the conclusion that the geometry of spiders' webs just wasn't a suitable tool for diagnosing mental illnesses.

But the researchers did find out one thing: namely, that the concentrated urine 'must taste extremely unpleasant, despite all the sugar that was added'. The spiders' behaviour left no room for doubt: 'After taking just a sip, the spiders exhibited a marked abhorrence for any further contact with this solution; they left the web, rubbed any residual drops off on the wooden frame, only returned to the web after having given their pedipalps and mouthparts a thorough cleaning, and could scarcely be persuaded to take another drop of the stuff.'

1955 The Psychonaut's Bathtub

If you watched the 1980 film *Altered States* right to the end, then the standard disclaimer in the closing credits must have struck you as somewhat unnecessary: 'The story, all names, characters and incidents in this production are fictitious. No identification with actual persons, places, buildings and products is intended or should be inferred.'

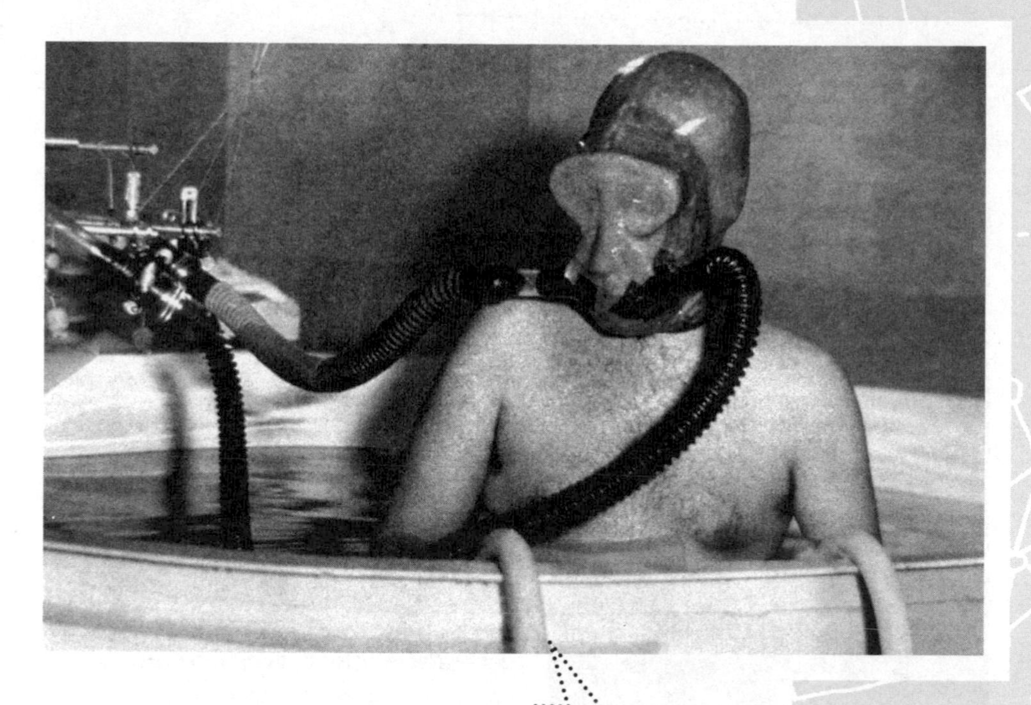

How does the brain react to sensory deprivation? A test subject wearing a breathing mask in the isolation tank.

As if anyone would even dream of taking this movie's convoluted plot to be a true story. In the film, the scientist Eddie Jessup (William Hurt) begins researching different states of consciousness in an isolation tank. However, the situation gets out of hand. His wife only just manages to save him from being assimilated into a swirling mass of primordial energy. He subsequently travels back into his evolutionary past, where he is transformed into an early hominid.

But however bizarre the plot might sound, the disclaimer is in fact untrue – *Altered States* is actually based on the experiments of the physician John Lilly, who credited the film's director Ken Russell with having done a 'good job'. The author of the novel on which the film was based, Paddy Chayefsky, took as his source material excerpts from Lilly's biography *Dyadic Cyclone* – for instance, the scene in which Jessup is rescued by his wife. In addition, the business with the early hominid did actually happen, albeit not to Lilly himself but to one of his colleagues who was experimenting with drugs in the flotation tank. 'Dr Craig Enright … suddenly "became" a chimp, jumping up and down and hollering for twenty-five minutes … I asked him later, "Where the hell were you?" He said, "I became a prehominid, and I was in a tree. A leopard was trying to get me. So I was trying to scare him away."

The plot of the film *Altered States* (1980) was based on John Lilly's experiments on sensory deprivation.

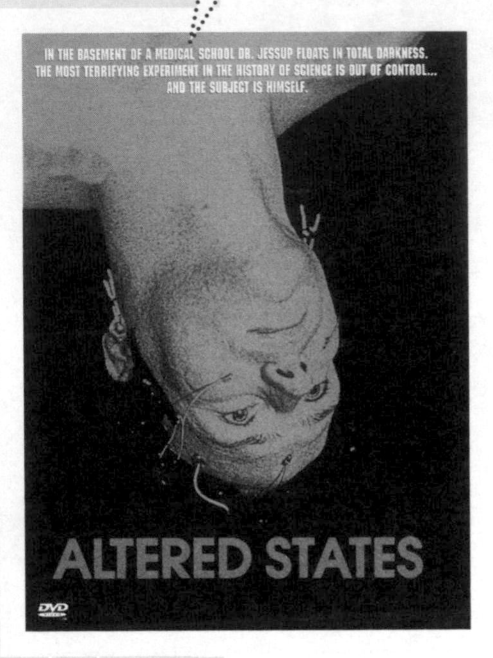

For a good portion of his life, there was nothing to suggest that John Lilly would one day spend his time talking to dolphins, meeting aliens, or discovering the Earth Coincidence Control Office. Lilly was a brilliant scientist with degrees in biology, physics and medicine when, in 1954, he began investigating a familiar problem in neurophysiology: namely, what would happen if the brain was cut off from all external stimuli? When neither a person's eyes nor their ears, skin or nose could register a single thing? There were two schools of thought on this question. The first was that the brain would shut down and the subject fall into a coma. The second reckoned that the brain generates its own activity and contains some form of internal pacemaker that keeps it awake, even in the absence of external inputs (see p. 135).

To test this hypothesis, Lilly set up his first sensory deprivation tank in an abandoned building in the grounds of the National Institute of Health in Bethesda, Maryland. It consisted of an outsized bathtub filled with warm water at precisely 34.5°C and situated in a soundproofed room that could be blacked out. Once a person was in the tank, all external distractions, including gravity, were reduced to an absolute minimum.

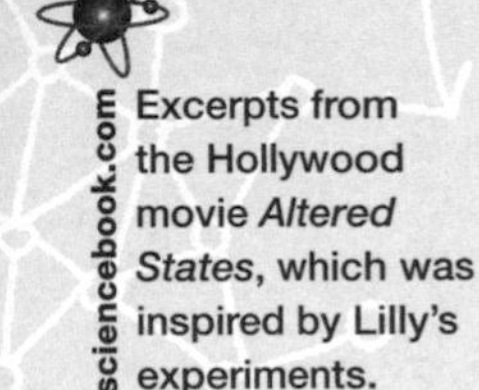

madsciencebook.com

Excerpts from the Hollywood movie *Altered States*, which was inspired by Lilly's experiments.

It took him almost a year to overcome all the tank's teething problems. In particular, he took ages to come up with a comfortable breathing mask. The test subject's head was underwater when he lay on his back in the tank. The rubber breathing mask covered the mouth, nose and ears, and was equipped with two snorkels around the mouth region. Anyone wearing it looked like a monster. Because the subject's legs weren't sufficiently buoyant, his feet rested on a rubber strip – the sole contact he had with anything other than water.

But by the end of 1954, everything was running like clockwork. The only slightly tricky element remaining was the business of getting into the tank. Lilly often worked on his own: after donning the closed mask, he had to climb the ladder blind, switch off the light and then let his body slip into the water, all the while trusting that he wasn't going to drown. Once in the tank, however, the duration of his stay was only limited by the need to take meals or keep appointments in the 'outside world' – his term for everything that took place outside the tank. He could urinate in the tank, as the water was changed on a regular basis.

A year later, Lilly published his first scientific paper on the sensory deprivation tank. In it he described the experiences of people who had spent time in total isolation, such as polar explorers and those who had been shipwrecked, and compared them with the various phases he went through during his isolation in the tank. The first three-quarters of an hour were dominated by thoughts of his day-to-day existence: Lilly was fully aware of where he was and mulled over things that had happened in the previous few days. Then he started to relax and enjoy the feeling of not having to do anything. But in the course of the next hour, he began to crave external stimuli. The researcher made lazy swimming movements so as to feel the water around him and flexed his muscles. His whole attention was focused on the few things he could still feel, i.e. the breathing mask and the rubber strip under his feet.

John Lilly experienced hallucinations in the sensory deprivation tank. He later gave up scientific research and became something of a New Age guru.

Having endured this phase without leaving the tank, he began to experience vivid fantasies. 'These are too personal to relate publicly,' he reported. He then entered the final phase, in which he started projecting images. On one occasion, the dark curtain before his eyes drew back after two and a half hours in the tank. Strange objects with shining outlines loomed up, and he saw a shimmering blue tunnel. Lilly could have kept staring at these apparitions for hours if his breathing mask hadn't sprung a leak, causing him to cut short the experiment. He had become a psychonaut on a journey of discovery into the self.

So, it was clear that the brain didn't lapse into a coma when it had no sensory input; in fact, quite the opposite – it was brilliant at creating its own entertainment. Yet answering that particular question had long since ceased to be the focus of Lilly's interest. He gave lectures at symposia held by widely differing fields of study, delivering one, for instance, to the American Psychiatric Association on the subject of 'Research techniques in schizophrenia', and another at a symposium entitled 'Psychophysiological Aspects of Space Flight'. The US military began to take an interest in Lilly's work, as it was well known that China and Korea had brainwashed prisoners by keeping them in solitary confinement during the Korean War.

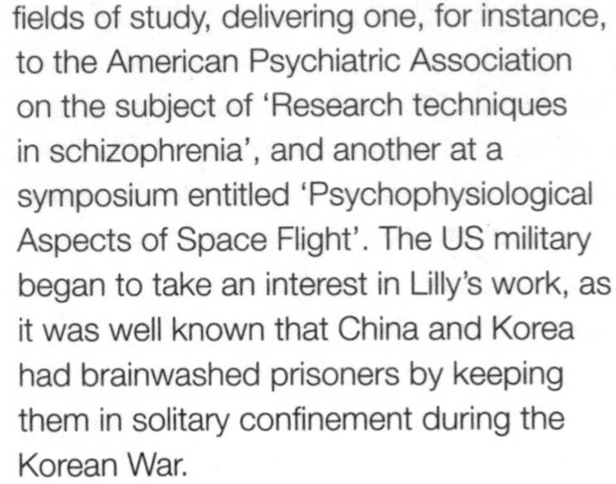

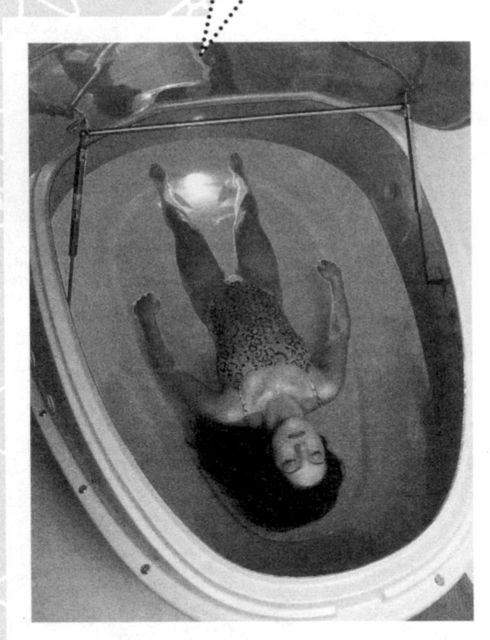

In the 1980s Lilly had isolation tanks (also known as Samadhi tanks) manufactured and marketed as relaxation aids. Modern versions are now frequently found in health clubs. Pictured below is an OVA model made by Jürgen Tapprich.

His experiences in the tank appear to have left a deep impression on Lilly. As he explained, 'I found many things there I didn't dare write about at the time because I was in the National Institute of Mental Health as one of the researchers rather than one of the patients.' He later linked his experiences in the tank to thought transference, and developed a theory of the human mind that 'would place much of the intellectual foundations of psychiatry on dangerously shifting sand', and compared his situation to that of Einstein when he began to realize the true potential of splitting the atom.

These weren't the kind of things that scientific journals were in the habit of publishing. So Lilly gave up his post at the NIMH and moved first to the Virgin Islands to study human communication with dolphins, and then to Miami, Baltimore, Malibu and Chile. All the while, he continued to conduct experiments in the tank. On the Virgin Islands he filled it with heated seawater and discovered that he could float in it virtually unaided and could dispense with both the stabilizing rubber strip and the unwieldy breathing mask. He tried using ordinary table salt dissolved in water. Because this made the slightest cut on the skin sting like fury, Lilly rubbed silicone gel over his whole body before stepping into the water. In later experiments he tipped sackfuls of the less astringent substance magnesium sulphate into a tank measuring 7 ft (2 m) in length, 3½ ft (1 m) wide and 10 inches (25 cm) deep. Even the most wiry of subjects found it easy to float in this.

The firm Samadhi Tanks, founded in 1972, started manufacturing tanks built to Lilly's specifications for domestic use, which quickly became popular in New Age circles. In Sanskrit the word samadhi means a state of deep inner calm.

In the 1980s it became the vogue, at around $30 an hour, for business executives to spend their lunch hours floating in isolation tanks. This activity, they variously believed, would help boost their self-confidence, sharpen their creativity, enable them to experience time travel, or generate good-luck hormones. Indeed, to this day, Samadhi Centres still sell themselves on the basis of all these supposed benefits and then some. However, their heyday has passed, and not all their customers find that the dark curtain opens for them or that they are party to fantasies that are too personal to be aired in public.

Because the experiences in the tank were peculiar to individuals, this made them impossible to pin down in any scientific way. Lilly was primarily interested in his own reactions. In actual fact, it quite suited him that the sensory deprivation tank had, as a result of its loose association with brainwashing, gained a certain notoriety, making it difficult for him to recruit test subjects.

In the 1960s Lilly's colleague Jay Shurley, who established his own research programme in this field at the University of Oklahoma, tried to undertake a systematic investigation of the effects of the sensory deprivation tank. But he soon found himself at a complete loss as how to categorize the visual fantasies of his test subjects scientifically. 'For example,' he wrote, 'consider how one would classify this: "I strongly felt that I was stirring with my right leg, and it was a spoon in an iced tea glass, just going round and round."'

Lilly became a guru of the New Age movement, wrote several zany volumes of autobiography, and combined his spells in the tank with acid trips. A woman journalist who asked former colleagues of Lilly in the 1980s where she could find him got the answer: 'Do you mean, what dimension?'

Lilly died of heart failure on 30 September 2001. He was 86 years old.

Source

Lilly, J.C. (1956), 'Mental Effects of Reduction of Ordinary Levels of Physical Stimuli on Intact Healthy Persons'. In Gottlieb, J.S., ed., *Research Techniques in Schizophrenia* (*Psychiatric Research Reports* 5), pp. 1–9. American Psychiatric Association.

1955 The Fog of Terror

That night, when the siren failed to sound a second time, Lloyd Long knew that it was going to be for real. Six days earlier, the 18-year-old had arrived with a group of volunteers at the Dugway Proving Ground in the Utah desert. And since then, the procedure had been the same every evening. Shortly before sunset, he and the other men would be picked up by a truck and driven to a remote part of the desert. There, they washed themselves under temporary open-air showers, put on clean clothes and, clutching blankets under their arms, took up their designated positions, sitting down in seats resembling bar stools that stood ready on the hard-packed sand in a long line stretching for almost a kilometre. On raised platforms between the stools stood cages containing rhesus monkeys and guinea pigs.

Whenever the siren sounded, Lloyd Long was required to look in the direction of Granite Peak and breathe steadily. Colonel William Tigertt, the

The 'Eight Ball' at the Fort Detrick research facility in Maryland. This hollow sphere was used to test the effects of germ warfare on humans. It has since been placed on the National Register of Historic Places.

US Army doctor conducting the experiment, gave the participants the following advice: 'Remember, when you hear the vacuum pumps, breathe normally, just breathe normally.' Usually, the siren would then sound a second time, indicating that the experiment had had to be abandoned because the wind direction wasn't right. The men would then change back into their normal clothes and be taken back to the barracks.

But on this night, 12 July 1955, wind conditions were perfect. A light breeze blew down from Granite Peak and Lloyd Long could hear the pumps which, one kilometre away from where he was sitting, started to spray a single litre of bacteriological agent into the night air. The agent that had been chosen for this test was the pathogen that produced Q fever, which brought on severe headaches, muscular pains and a high fever. In most cases, the disease ran its course with no lasting ill effects, but at that time around 1 in every 30 people who contracted it died. Lloyd Long's group consisted of 30 people.

Long scarcely registered the fine mist that drifted over him. It was only when men in protective clothing showed up that he knew the experiment was over. He was required to shower, stand under an ultraviolet lamp that would kill any residual microbes, and then shower once again. His clothes were burned. Thereafter, the whole group was flown to Fort Detrick near Washington DC. The first – and, according to US Army records, to date sole – experiment conducted on human beings with biological weapons released into the open air was entering its second phase.

The US government at the time knew that the Japanese had conducted extensive experiments with biological weapons during the Second World War, and assumed that the Russians were also undertaking similar tests. Although the USA officially condemned the manufacture and use of biological weapons, they secretly launched their own research programme into producing them in 1943. Fort Detrick served as the headquarters for the scientists engaged on this programme. Animal experiments carried out there tested the suitability of certain pathogens for use in weapons, while at the same time the scientists developed vaccines for their own forces. Yet it was always difficult to extrapolate from animals to man. As one document prepared by the US Air Force noted, 'The Air Force could be fairly accurate in predicting what a biological warfare attack would do to a city full of monkeys, but what an attack would do to a city full of human beings remained the 64-dollar question.' The military came to the conclusion that only tests on humans would succeed in getting them any further.

It's not unheard of for humans to be used as guinea pigs in medical experiments. Before it is introduced to the general public, every pharmaceutical is taken by test subjects. However, the tests with the Q fever pathogen differed inasmuch as this wasn't an attempt to cure people with a drug, but rather to make them ill with a pathogen. But as long as the participants gave their assent to the tests, then the people at Fort Detrick could see no problem with it.

In the US Army there was a particular group of soldiers who were ideally suited to this task. Men who belonged to the Seventh-Day Adventist community were exempted on religious grounds from having to fire or handle any weapons, but were extraordinarily healthy – they didn't smoke and drank neither alcohol nor coffee. In addition, many of them were vegetarians. 'You didn't have to ask if their reaction was because they were drunk as a skunk on Saturday night,' as one churchman put it in explaining the advantages of a sober lifestyle for medical research.

Colonel Tigertt got in touch with the elders of the Seventh-Day Adventist Church, who were soon persuaded of the honourable nature of the experiment and officially approved the plan to recruit Adventists to take part in the tests. On 19 November 1954 Theodore Flaiz, general secretary of the Adventist Church and its spokesman on medical matters, reacted enthusiastically to a government request: 'The type of voluntary service which is being offered to our boys in this research program offers an excellent opportunity for these young men to render a service which will be of value not only to military medicine but also to public health in general.' Between 1955 and 1973, 2200 young men volunteered for this detail. The codename for the 153 top-secret experiments – which included infecting participants with anthrax, tularaemia, typhoid fever and meningitis – was 'Operation Whitecoat'.

The first tests were the Q fever experiments that Lloyd Long took part in. Yet even before the open-air experiments at the proving ground in Utah, the 'Eight Ball' at Fort Detrick had come into operation: this was a hollow sphere, 40 ft (13 m) high, made of stainless steel, which the scientists at the facility had named after the black ball on a pool table. Whenever tests were due to be conducted, Adventists would step into the telephone-booth-sized chambers attached to the side of the 'Eight Ball'. Inside, they donned breathing masks that were connected to the inside of the sphere. A technician would then use a remote-control device to release a fine spray of bacteria or viruses inside the sphere. The men breathed in the mixture for one minute and were then immediately taken to a medical ward, where they were placed in isolation, under observation.

The same procedure was adopted for the men who returned from Utah: in their single rooms, equipped with television, books and games, the participants waited for the thumping headaches to start, signalling the onset of Q fever. Around one-third of the men actually came down with the disease. The severity of the symptoms depended on whether they had acquired any immunity from taking part in earlier experiments involving the 'Eight Ball'. Precisely where their stools were placed in the desert tests was also a decisive factor. Lloyd Long, who was seated on the edge of the infected area, was up and about again after just a day in bed. All the test subjects made a full recovery.

Today, most of the Whitecoat veterans are proud of their involvement in the programme. 'I don't know anyone who went in who felt later on that they had been bamboozled,' claims Lloyd Long, who is now in his late sixties and a retired insurance salesman. Some of the Whitecoat veterans have become regular television interviewees following the attack on the World Trade Center and the growing climate of fear surrounding bio-terrorism. However, some criticism has been voiced about the close connection between the military and the Adventist church, and particularly in the 1960s questions were asked about whether it was right for a church that advocated non-violence to be supporting a germ warfare programme. But in the light of some of the other things that were going on in the same period, scrutineers of ethics have given Operation Whitecoat a pretty clean bill of health. The volunteers were informed repeatedly about the potential risks, and were free to quit the programme any time they chose. Even so, it's hard to imagine experiments of this kind being sanctioned nowadays. The risk for such a sensitive organ as the lung is simply too great.

The bacteria that were released in the desert were killed off by exposure to sunlight on the day after the test. Moreover, not one of the guinea pigs that were also placed alongside Highway 40 – some 34 miles (55 km) from the proving ground – to test the effects of the bacteria, ever suffered any ill effects.

Source

Regis, E. (1999), *The Biology of Doom: The History of America's Secret Germ Warfare Project.* Holt, pp. 172–176.

1957 Psychology's Atom Bomb

At a press conference in New York on 12 September 1957, the American market researcher James Vicary unleashed a certain form of paranoia that still haunts some people even today. Vicary showed the assembled journalists a short film about fish. During the screening, a special projector kept flashing up the instruction 'Drink Coke!' – as often as every five seconds on some occasions. Each image lasted for only 1/3000th of a second – too short a time, in other words, for the newspapermen to pick up the message consciously. Only when the lecturer deliberately made the image darker could the messages be seen clearly imprinted on the film like a watermark.

Vicary claimed that he had also carried out the experiment shortly before in a cinema in Fort Lee, New Jersey. Over a period of six weeks, a total of 45,699 people who went to the movie theatre there were unwittingly exposed to

In this cinema in Fort Lee, New Jersey, filmgoers were allegedly manipulated by means of subliminal messages.

the secret instructions 'Eat Popcorn!' and 'Drink Coke!' With the result that the sales of Coca-Cola in the cinema foyer increased by 18.1 per cent, while those of popcorn rose by 57.5 per cent.

Huxley Fears New Persuasion Methods Could Subvert Democratic Procedures

The author Aldous Huxley puts his oar into the debate on subliminal advertising (from the *New York Times*, 19 March 1958).

Public opinion was outraged at this. If a person was able to implant in cinemagoers' brains the urge to eat popcorn without their knowledge, then what was to stop him ordering them to commit a murder? Or to send a battalion of brainwashed zombies off to war? Or, for that matter, to suggest to women that they should give up vacuuming?

'Minds had been broken and entered,' wrote the *New Yorker*. The novelist Aldous Huxley warned of the 'alarming danger' that people might lose control over their own minds, just as he had predicted in his novel *Brave New World*. Moreover, the Christian organization Women for Abstinence raised the spectre of these diabolical advertising slogans being used by breweries and distilleries to boost their business. Only the fashion magazine *Vogue* found anything positive to say about the matter. It rolled out a 'subliminal dress' made of black crêpe de Chine and costing $160, which achieved its effect by 'tapping out its message to the subconscious'.

Yet Vicary was by no means the only person at that time who was experimenting with subliminal messages. Psychology had long been interested in what effect information had below the threshold of conscious perception. But he was the first person to announce that he had manipulated cinemagoers by remote control. Within a month, his company Subliminal Projections, as he explained at the press conference, had plans to equip 15 cinemas with the special projector for a three-month test period.

Vicary claimed that his hidden advertising would finally free television viewers from tiresome commercial breaks: 'Many nights I've tried to watch the late night show movie on TV, but just before John kisses Mary, some sewer-cleaning commercial interrupts the show.' The subliminal messages, he reckoned, would be a positive boon for the public.

The trouble was, the public saw things quite differently. The journalist and critic of advertising Vance Packard had recently, in his book *The Secret Seducers*, revealed what tricks the advertising industry used to influence people's purchasing decisions. The book became a bestseller, and Vicary's experiment seemed to bear out Packard's dire warnings. In no time the market researcher's special projector became known as the 'atom bomb of psychology'. It was evidently high time that politics intervened.

After senators had raised the matter several times in Congress, Vicary travelled to Washington in January 1958 to demonstrate his new marketing technique to America's politicians. According to the report in the advertising industry's trade journal *Printer's Ink*, there was something quite grotesquely perverse about the film presentation with the hidden popcorn message: 'Having gone to see something that is not supposed to be seen, and not having seen it, as forecast, the FCC [Federal Communications Commission] and Congressmen seemed satisfied.' *The New York Times*, on the other hand, remarked that some of the politicians were disappointed not to have felt the urge to buy popcorn. The only direct reaction that was recorded for posterity was that of the Republican senator Charles E. Potter, who is reported to have announced mid-screening, 'I think I want a hotdog.'

madsciencebook.com

Party political broadcast by the Republican Party during the US presidential elections of 2004, in which the word 'Rats' was briefly flashed up.

Vicary had a ready explanation for the apparent failure of his advertising technique: 'Those who have needs in relation to the message will be those who respond.' Subliminal advertising was, he claimed, a very mild form of the art; it would never, for instance, persuade a Republican to become a Democrat.

After the demonstration in Washington, it began to become apparent that Vicary's claims weren't entirely kosher. All attempts to repeat his experiments failed, and scientists gradually started to lose their patience with the dodgy market researcher, who, citing the fact that the patent was still pending, persisted in refusing to reveal the exact procedure and the precise data relating to his experiment. Rumours began to circulate at this time that Vicary had pocketed $4.5 million in consultant's fees from advertising agencies. If this was true, then it was a waste of money. A trip to Fort Lee, where the first experiment was alleged to have taken place, would have shown straight away that the small cinema there couldn't possibly have had 45,699 visitors through its doors in the space of six weeks.

In 1962, in the trade journal *Advertising Age*, James Vicary finally admitted more or less openly that the whole story had been fabricated. Sure, his projector worked all right, but the method clearly had no measurable effect: 'We applied for a patent after testing the thing in a movie theatre in Fort Lee, N.J. The story leaked out to some newspaper guys and we were forced to come out with subliminal before we were really ready … I had only a small amount of data – too small to be meaningful.'

Vicary later disappeared without a trace. Nobody knows whether he's still alive or whether the last version of events that he offered was in fact true.

The supposed experiment still lives on in modern times as something of an urban legend. The producers of self-help cassettes containing subliminal messages aimed at getting people to lose weight or boost their self-confidence continue to cite it as incontrovertible proof of the efficacy of their product. It has also entered popular culture. There are a number of films whose plot is based on the premise that it's possible to influence people subliminally. Even Inspector Columbo, in the 1973 episode 'Double Exposure', solved a crime involving a dubious market researcher by using subliminal messages.

Meanwhile, the study of subliminal perception has become a blossoming branch of scientific research. Nowadays, simple experiments can be used to demonstrate whether people have taken in information that they're not aware of, and whether this information has influenced their actions. However, the effect has been shown to be slight and certainly doesn't lead, say, to an increase of almost 60 per cent in popcorn turnover.

Vicary's experiment had its last major airing to date during the US presidential elections of 2000, when campaign managers working for the Democratic contender Al Gore noticed something odd in a TV advert promoting the Republican candidate George W. Bush. Unseen by viewers, the word 'Rats' was flashed up momentarily across the whole screen when a Democrat policy was mentioned. The producer of the film defended his actions with the wounded assertion that the insertion had only been intended to highlight the word that followed –'Bureaucrats' – and make it more eye-catching. But it's more likely that the rats in the Republican Party political broadcast were late heirs to Vicary's experiment – an experiment that never actually took place.

Source

Talese, G. (1958), 'Most Hidden Persuasion'. *New York Times Magazine*, 12 January, p. 22.

1958 The Mother-Machine

Paradoxically, some of the most upsetting and cruel experiments in the history of science were conducted in an attempt to research the nature of love. They were devised by Harry Harlow, a workaholic psychologist who was also an alcoholic, a difficult husband and a distant father. His findings on love changed the business of bringing up a child for ever.

Harlow's research focused on the most fundamental form of affection: motherly love. He lighted on this area of study in attempting to breed rhesus monkeys for his experiments on learning. To protect them from diseases, he separated the monkeys from their mothers shortly after birth and raised them instead by bottle-feeding them in individual cages. The animals were healthier and heavier than those who had grown up in a natural environment, which convinced Harlow that he was actually a better mother for the monkeys.

The psychologist Harry Harlow with his infamous invention: the terry-cloth mother.

But despite the fact that the baby monkeys apparently wanted for nothing, they sat hunched in their cages, sucking their fingers and staring aimlessly into the middle distance. When Harlow later paired males and females, the animals didn't have the faintest idea what they were supposed to do with one another. This surprised him, since a basic principle of science at that time was that babies, in order to have the best possible chance of developing normally, needed first and foremost enough to eat and to be kept clean. And indeed, both these criteria had been met in the case of his monkeys.

From psychology's point of view, motherly love was a second-rank emotion, which only came into play once the mother had satisfied the far more important requirements of her offspring, namely hunger and thirst. This was also the case for people. Experts on child-rearing advised parents against cuddling their children. The psychologist John B. Watson (see p. 79) led a crusade against the evils of excessive parental love. In his bestselling book *Psychological Care of Infant and Child*, which appeared in 1928, there is a chapter entitled 'The Dangers of Too Much Mother Love'. In it, Watson contended that too much affection lavished on a child would inevitably lead to problems for that individual in adulthood. If you absolutely must kiss your child, he said, then do so only on the forehead.

Even though the wire-mesh mother provided milk, the baby monkeys preferred the terry-cloth mother.

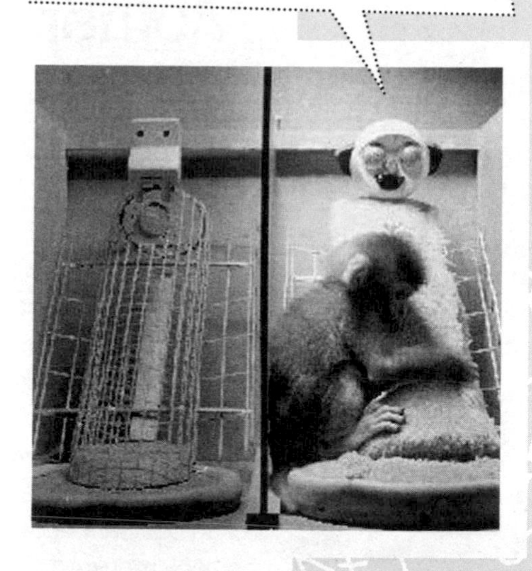

Alongside apathy, Harlow's baby monkeys also exhibited another unusual behaviour – a fanatical attachment to the fabric linings of their cages. The monkeys would cling on to them, wrap themselves up in them, and started shrieking when they were changed during the regular cleaning of the cage. If they spent the first five days in the cage without such a piece of material, they barely survived. Could it be that this soft cloth was just as important as the bottled milk they were fed?

Harlow tried an experiment: he built a mother for his little monkeys. Its head was made out of a maple-wood billiard ball, while its eyes were bicycle reflectors. But this part of the model wasn't really important. The essential bit was the cylindrical body, comprising a piece of towelling wrapped around a small fluffy cushion. Next to this terry-cloth mother he placed a second one shaped exactly the same but made of wire mesh with no soft padding but with a bottle of milk at chest level. If the prevailing scientific view was correct, reasoned Harlow, then the baby monkeys would show a strong preference for the wire mother, because she provided them with their only means of sustenance. But the opposite turned out to be the case: the monkeys clung on to the terry-cloth mother for more than 12 hours a day, and only climbed onto the wire mother briefly when they were thirsty. Harlow thereby proved that a baby's affection is focused primarily on the mother's soft, warm body, irrespective of whether it is also the source of food. In other words, he showed how vital bodily contact is for a child's development.

madsciencebook.com

Old TV footage of Harry Harlow describing his experiments.

The Parent Problem—
Mother-Machine Works

Headline from the *Stevens Point Daily Journal*, 15 November 1958.

The experiment with the terry-cloth mother was just the start of an extensive research programme that Harlow conducted on love and on what happens when baby monkeys receive none. For his next test, he constructed monster mothers, fluffy like the original terry-cloth mother but at the same time deceitful and cruel. One was designed to shake the baby off repeatedly, while another frightened it by sending out blasts of compressed air, while yet another was equipped with hidden metal spikes that suddenly flicked out of its body and dislodged the infant. And what did the baby monkeys do? As soon as the mother had 'calmed down', they went back to her and nuzzled up close to her. This happened time and time again. Harlow's monster mothers were an impressive demonstration of a baby's craving for its mother and its total dependency upon her.

Even crueller was the 'pit of despair', a funnel-shaped cage at the lowest point of which the baby monkey was placed. For the first two or three days it would try in vain to clamber up the steep sides of the cage. Then it would give up and simply stay sitting there, lonely and despairing. Within a short space of time, the monkey became what we would call in human terms depressive, and Harlow would then try to cure it with drugs or by introducing it to other monkeys, which worked in some cases.

Harlow never denied that the monkeys were subjected to suffering in his experiments. Nor did he ever express regret. In fact, he once told a newspaper reporter: 'Remember, for every mistreated monkey, there are a million mistreated children. If my work will point this out, and save only one million human children, then I can't get overly concerned about ten monkeys.'

Ironically, Harlow never cared much for his own children. His first wife left him and took them with her, since, as she claimed, living with him was tantamount to living alone anyway. His second wife died of cancer when he was 66. Eight months later he got remarried – to his first wife.

madsciencebook.com

In 2002 Pulitzer prizewinner Deborah Blum published a fascinating biography of Harry Harlow: *Love at Goon Park: Harry Harlow and the Science of Affection* (Perseus Publishing).

Source

Harlow, H.F. (1958), 'The Nature of Love'. *American Psychologist* 13, pp. 573–685.

1959 Fouling up the Fighter

Eight years had passed since pilots had shown that they became weightless for a short time when they flew in a parabola (see p. 132). In 1951 the scientists' main concern had been to find out how a plane's occupants got their bearings in such circumstances. But in the meantime, things had changed dramatically. The Soviet Union and the USA had launched their first satellites into terrestrial orbit, and most people believed that the day wasn't far off when manned spacecraft would follow satellites into space.

But how would people cope with weightlessness that lasted far longer? While a person was in this state, would he even be able to satisfy his most basic physical needs, such as eating, drinking and sleeping? Eating didn't seem to present a problem. But as far as drinking was concerned, Captain Julian E. Ward and some fellow officers from the US Air Force set out to investigate just how practical this might be.

Ward instructed 25 volunteers to drink from a variety of different receptacles during the period of weightlessness during parabolic flight. These drinking vessels included an open beaker without a straw, an open beaker with a straw, and a plastic squeezy bottle that could squirt liquid directly into the mouth. The result was one huge mess. As Ward reported, 'Although it was realized that such attempts would be exceedingly messy, many of the ramifications had not been anticipated.' All but two of the test pilots had problems with the open beaker; as soon as they moved it, however slightly, an amoeba-like mass of water would fly out of it and cover their faces. When they tried to breathe, the water would enter their windpipes through their noses and trigger a coughing fit. The researchers were astonished to discover that a person could actually drown in this bizarre fashion. The straw was completely useless as well. The squeezy bottle, on the other hand, presented no problems whatsoever, at least not until just after the pilots had swallowed. At this point, several of them exhibited what Ward called the 'weightlessness regurgitation problem'.

A pilot attempting to drink during conditions of weightlessness.

Source

Ward, J.E. (1959), 'Physiologic Response to Subgravity I. Mechanics of Nourishment and Deglutition of Solids and Liquids'. *Journal of Aviation Medicine* 30, pp. 151–154.

Just light pressure on their abdomens was all it took to make their stomach contents move up to their heads and back into their mouths.

Investigating sleep during weightlessness turned out to be more complicated than these drinking experiments. The periods of weightlessness during parabolic flight lasted 30 seconds at most. On the other hand, the question wasn't necessarily whether sleep was possible in the absence of gravity, but rather what happens when a person wakes up. Some aviation medics took the view that people would be completely disorientated in such circumstances.

Lieutenant Clifton M. McClure set about finding whether this was in fact the case. After staying awake for 48 hours, he ate a hearty breakfast, which made him even more drowsy, and then got into the back seat of an F-94C Starfire jet. At a height of 11,500 ft (3500 m) he took off his headphones and fell asleep 25 minutes later. The pilot of the aircraft then flew a weightless parabola, during which he woke McClure by pulling on a cord that was tied to his left wrist. McClure's first impression was that his arms and legs 'were floating away from him'. He tried to cling onto the cockpit canopy, and was completely disorientated.

It later transpired that this phenomenon didn't really present a major problem. Far more serious was the effect of the decrease in pressure on a person's skeleton and muscles during weightlessness. This phenomenon was studied in an experiment in which test subjects lay in bed and did nothing – for a whole year (see p. 268).

1959 The Experiment with the Unabomber

On 3 April 1996, 150 heavily armed FBI agents stormed an isolated shack in a wood near the town of Lincoln in the state of Montana. Its occupant was a 54-year-old former mathematics professor with a first degree from the prestigious university of Harvard and a prizewinning doctoral thesis. These distinctions later earned him the title of the 'most intellectual serial killer in the USA'.

Between 1976 and 1995, Ted Kaczynski set off 16 home-made bombs, which succeeded in killing three people and injuring eleven others, some of them seriously. His attacks were meant as a protest against scientific and technological progress, which in his view was inexorably compromising the freedom of the individual. The FBI dubbed him the 'Unabomber' because his first victims worked for UNiversities and Airlines.

Following his arrest, people started to ask themselves why a brilliant mathematician, who at one stage could have had a glittering career ahead of him, should end up living in a shack with no electricity or running water and making bombs there. In his 2003 book *Harvard and the Unabomber*, the historian Alston Chase reckoned that he had found the answer to this question in the shape of the 'Murray Experiment'.

The Unabomber being taken into custody in 1996. But did a psychological experiment turn Ted Kaczynski into a serial killer?

From the beginning of 1960 onwards, Kaczynski took part in a three-year-long experiment being conducted by Henry A. Murray. Murray was a Professor at the Department for Social Relations at Harvard University, but by the time he made the acquaintance of Kaczynski as one of his test subjects, he was nearing the end of his career. He was 62 years old, had been responsible for developing a groundbreaking psychological test (the TAT, or Thematic Apperception Test), written a widely read book about it, and evaluated army recruits for their suitability for deployment on secret operations.

It's unclear how Kaczynski got wind of the experiment. It's possible that he saw the advertisement for volunteers: 'Would you be willing to contribute to the solution of certain psychological problems (parts of an on-going program of research in the development of personality) by serving as a subject in a series of experiments or taking a number of tests (average about two hours a week) through the academic year (at the current College rate per hour)?' On the other hand, it could be that Murray selected Kaczynski personally. For his experiment, he was looking for Harvard students in their first semester who had as diverse a range of personalities as possible, including self-confident people, conformists and insecure types. According to psychological tests, out of the 24 young men he selected for the experiment, Kaczynski was the most unstable.

So as to protect the private lives of the participants, Murray gave each of the students a codename. He called Kaczynski 'Lawful'. This may now seem deeply ironic, but in fact was not inappropriate: Kaczynski was unassuming and by no means a rebel. As the son of a working-class family, he felt ill at ease at Harvard, had only a small circle of friends and was weighed down by the great expectations of his parents. He worked hard and rarely went out.

The psychologist Henry A. Murray conducted controversial experiments, in which the test subjects' convictions were systematically undermined.

Murray called the core of the experiment the 'Dyad', which consisted of a stress-inducing debate. For this, the subject was required to sit in a brightly lit room in front of a one-way mirror, through which he was observed and filmed. Measuring devices recorded his rate of heartbeat and respiration.

Murray told all the participants that another student would strike up a discussion with them. But what he didn't reveal to them was that their interlocutor was a highly eloquent law student whom Murray had specifically trained to get under the skin of his test subjects. He was instructed to handle them roughly and ridicule their philosophy of life. Murray had gleaned the information on his victims' outlook on life from a series of tests and personal statements that formed part of the experiment. During the debate, all the subjects started off by trying to defend their attitude but found themselves having to give ground in the face of the virtuosic and cynical arguments put forward by their opponent. Eventually, they were all overcome with impotent rage.

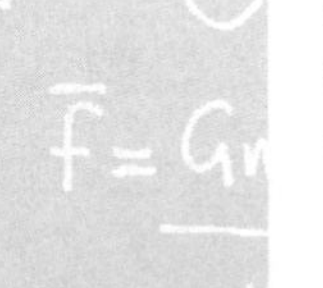

After this confrontation came a deluge of further tests and

discussions. One of these entailed the subjects viewing a recording of their argument and being asked to comment on their angry reactions.

To date, nobody quite knows what Murray was trying to achieve with this experiment. His aims were somewhat vague and confused. For example, he claimed he wanted to 'develop a theory of dyadic systems' and to use the data he collected to further people's personal development. But even his assistants had no real idea what the experiment was driving at. Murray's biographer wrote that the psychologist was just trying to find out what happened when one person attacked another.

Alston Chase believes that Murray's experiment had a quite different root. Murray had got married at 23, and seven years later met Christiana Morgan, who was also married, and with whom he embarked on a turbulent lifelong affair. Some of his early collaborators take the view that Murray's experiments were nothing more than reruns of this relationship. Shortly before his death in 1988, Murray gave an indirect confirmation of this suspicion. 'I have been asked why Christiana and I started a separate dyad', he wrote, and put forward a number of reasons for this, including the following two: 'I had a wish to develop my theory in which two people (not just one personality) are incorporated into one system, a dyadic system,' and 'we wanted also to experiment with different types of combinations in play and work'. In other words, Murray appeared to be treating the relationship as though it were an experiment. Chase concludes from this that the arguments in the Murray Experiment represented his relationship with Christiana Morgan.

Kaczynski later recalled that the dyad was a 'highly unpleasant experience'. Could it possibly have represented a turning point in his life, though? Not in and of itself, says Chase, for other factors included Kaczynski's general lack of ethical focus at that time plus his brittle personality.

Kaczynski developed his technophobic view of the world in his final years at Harvard. He became convinced that technology and science were threatening people's freedom and increasingly controlling their thinking.

After graduating from Harvard, Ted Kaczynski wrote a brilliant doctoral thesis at the University of Michigan and in 1967 accepted a position as assistant professor at the University of California at Berkeley. Two years later, however, he left Berkeley and built himself the hut in

madsciencebook.com

The Unabomber Manifesto: 'Industrial Society and Its Future'.

Source

Chase, A. (2003), *Harvard and the Unabomber. The education of an American terrorist*, W.W. Norton.

the woods outside Lincoln, where he hatched his plans for his bombing campaign.

The Unabomber's ultimate undoing was the statement that he sent simultaneously to the *New York Times*, the *Washington Post* and *Penthouse* magazine on 24 June 1995: this was in the form of a treatise entitled 'Industrial Society and Its Future', which became widely known as the 'Unabomber Manifesto'. In it, he held out the prospect of stopping his attacks if the essay was published.

On 19 September 1995 the *Washington Post* printed the text of Kaczynski's manifesto over 56 pages of newsprint. Shortly thereafter, David Kaczynski went to the FBI and told them of his suspicion that his brother Ted might be the Unabomber. He had found that certain passages in the manifesto were repeated verbatim from letters that Ted had sent him some time ago.

Ted Kaczynski was sentenced to life imprisonment without the possibility of parole on 4 May 1998. After Alston Chase had voiced his suspicion, in the June 2000 edition of the journal *The Atlantic Monthly*, that the Murray Experiment might have had something to do with Kaczynski becoming a bomber, the Murray Research Center – the institute at Harvard named after the psychologist – issued a firm rebuttal. Other students who had taken part in the experiment, claimed the Institute, had not found the argument stressful; moreover, Chase had misconstrued the purpose of Murray's work.

After Ted Kaczynski's codename 'Lawful' was made public, the Murray Research Center placed an indefinite embargo on access to the raw data from the experiment.

1959 A Trinity of Christs

The unlikely encounter took place on 1 July 1959 on Ward D-23 at the State Psychiatric Clinic at Ypsilanti near Detroit, Michigan. The three men, whom the psychologist Milton Rokeach brought together in a small, plainly furnished visiting room at the institute, introduced themselves in turn. The first to do so was a 58-year-old with a bald head and gappy teeth.

'My name is Joseph Cassel. I'm God.'

Next up was a 70-year-old, whose mumbled introduction was hard to make out.

'My name is Clyde Benson. I made God.'

Finally, a 38-year-old man with an emaciated body and grave expression stepped forward, but refused to give his real name, Leon Gabor.

'It states on my birth certificate that I am the reincarnation of Jesus of Nazareth.'

Thus began one of the most bizarre experiments in the history of psychology. The proposition was to find out what happens when people are confronted with the most extreme paradox imaginable, namely with another person who claims to have the same identity. How would the three men react to the realization that, all of a sudden, there was more than one Jesus? (God and Jesus were identical as far as they were concerned.)

Three 'Christs' Together In State Hospital

EAST LANSING, Jan. 29 (AP)—Three mental patients — each claiming to be Jesus Christ—have been brought together at the Ypsilanti State Hospital.

Milton Rokeach's unusual study makes the headlines (*Herald Press*, 29 January 19.60).

Milton Rokeach had already spent a long time investigating the relationship between a person's identity and her innermost beliefs. What internal canons of behaviour are central to determining one's personality? Which of these can be altered without any consequences? And what happens when one of the main planks of a person's belief system comes under threat?

Rokeach had seen from the example of his own children how sensitive people were to any violation of their identity. One time, when he jokingly mixed his two daughters' names up, their laughter was soon replaced by unease. 'Daddy, this is a game, isn't it?' the younger daughter asked nervously. He answered no, it wasn't, and soon afterwards both girls were begging him to stop. Rokeach had attacked the very core of their innermost conviction, namely their sense of self.

Rokeach could only hazard a guess at what would have happened if he had kept on mixing their names up for a whole week. Clearly, conducting an experiment along these lines was out of the question on ethical grounds. Yet reports from Chinese prisons, where brainwashing was carried out using similar techniques, suggested that the effects on a person's identity were severe.

In casting around for an experiment that would give no cause for concern, Rokeach had a sudden flash of inspiration: why not use mental patients who think they are someone else? If he could bring together under one roof several of them who all claimed to be the same person, this would cause two fundamental beliefs to collide: their false conviction as to who they were and their correct conviction that two people can't have the same identity.

In psychological literature, Rokeach found two brief examples of such cases: in the 17th century, two men who both thought they were Jesus Christ met by chance in a lunatic asylum. Three hundred years later, also in a psychiatric institution, two 'Virgin Marys' also came face to face. In both cases, the meeting was said to have led to a partial recovery.

Rokeach hoped that the experiment would not only reveal more about people's internal belief system but also suggest new therapeutic possibilities for patients with severe personality disorders. He made enquiries at all five of the psychiatric facilities in the state of Michigan in his search for two psychotics who claimed to have the same identity. Among the 25,000 patients, there was only a handful of such cases. There were no Napoleons, no Khrushchevs and no Eisenhowers. There were just a few people who thought they were members of the Ford or Morgan dynasties, plus a female God, a Snow White and a dozen Christs.

Of the three men who thought they were Christ and who were suitable subjects for the experiment, two were resident at the clinic in Ypsilanti. The third one was transferred there. Over a period of two years, they slept in adjacent beds, ate at the same dining table and were assigned similar duties in the hospital laundry.

Leon Gabor had grown up in Detroit. His father had run off and left the family, while his mother was a religious fanatic. She spent the whole day praying in church, and left the children to fend for themselves at home. Gabor enrolled at a seminary for a short time before enlisting in the army. Later, he went back to live with his mother, who completely dominated him. In 1953, at the age of 32, he began to hear voices telling him that he was Jesus. One year later he fetched up in a psychiatric hospital.

Clyde Benson grew up in the Michigan countryside. When he was 24, his wife, his father-in-law and his parents all died. His eldest daughter married and moved away. Benson started drinking and remarried, lost everything he owned, became violent and eventually landed in jail, where he claimed to be Jesus Christ. In 1942, aged 53, he was referred to a psychiatric institution.

Joseph Cassel was born in the Canadian province of Quebec. He was something of a misanthrope, burying himself in his books and making his wife take a job to support him while he worked on writing his own book. He and his family moved in with his in-laws, where he lived in constant fear of being poisoned. It was this delusion that brought him to Ypsilanti in 1939. At the time, Cassel was 39 years old. Ten years later he started to believe that he was the Father, the Son and the Holy Ghost.

After just a few encounters, each of the three had a ready explanation for the fact that the other two claimed to be Jesus. Thus, Benson claimed: 'They are really not alive. The machines in them are talking. Take the machines out of them and they won't talk anything.' Meanwhile, Cassel's explanation was disarmingly logical: Gabor and Benson couldn't be Jesus because they were self-evidently patients in a psychiatric institution. Gabor had various explanations for the others' impossible identity. For example: they only made out they were Jesus to gain prestige. Even so, he did go so far as to concede that they might be 'hollowed-out instrumental gods with a small "g".'

To get to know the three men better, Rokeach set the topics for discussion at each of their daily sessions. They talked about families, their childhood, their wives and – repeatedly – about their own identity. Heated debates ensued, which after three weeks led to the first violent clash: when Gabor claimed that Adam was a negro, Benson clouted him. After two further physical altercations – between Benson and Cassel and Cassel and Gabor respectively – the three Jesuses conducted themselves peaceably for the rest of the experiment. However, they stuck to their guns on the question of who they believed themselves to be. Only Gabor, presumably influenced by the smack in the mouth Benson gave him, changed his mind about Adam, conceding that he might not, after all, have been black.

After two months, Rokeach let the three men lead the discussions. Each of them in turn chaired the daily meetings, chose the topic for discussion and handed out the daily cigarette ration. They covered a broad spectrum of subjects: films, communism and religion, for example, but never touched on the question of their own identity again. And if one of them just happened to mention in passing that he was God, the others deftly changed the subject.

Yet all this did nothing to shake the conviction of each of them that he was the real Christ. Gabor showed the hospital staff his handwritten visiting card, on which he had inscribed: 'Dr Domino dominorum et Rex rexarum, Simplis Christianus Puer Mentalis Doctor, reincarnation of Jesus Christ of Nazareth'.

Source

Rokeach, M. (1964), *The Three Christs of Ypsilanti*. Alfred A. Knopf.

However, in a surprise move in January 1960, about six months after the first meeting, Gabor changed his name. Now the visiting card read: 'Dr Righteous Idealed Dung Sir Simplis Christianus Puer Mentalis Doctor.'

'What do you want us to call you?' asked Rokeach.

'If you want to say Dr Dung, sir, that's your privilege,' replied Gabor.

This name caused some difficulties in the clinic. The nurses refused to call a patient 'Dung', but Gabor wouldn't answer to any other name. Finally, he and the matron settled on the name 'R.I.', from 'Righteous Idealed'.

Rokeach immediately asked himself whether the name change betokened a change of identity on Gabor's part. But in all likelihood his motivation was simply to remove himself from the firing line and remove any grounds for further confrontation.

In the course of the experiment, Rokeach deliberately intervened on several occasions in an attempt to learn more about what made the men tick. For instance, he proposed taking their stated identities at face value and differentiating between them by calling Cassel 'Mr God' and Benson 'Mr Christ'. The men turned this suggestion down. They were evidently well aware that no one except them shared their conviction, and that an official name change would only cause more problems. Another time he read them an article about the experiment from the local paper. Rokeach then asked Benson:

'Do you know who they are?'

'No, I don't," replied Benson.'

'Do you have any idea?'

'No, their names aren't in the article.'

'What about the one who's better?' asked Rokeach, meaning Gabor.'

'He is not wasting his time trying to be Jesus Christ.'

'Why is it a waste of time?'

Benson stuttered a little as he responded: 'Why should a man try to be somebody else, when he's not even himself? Why can't he be himself?'

Later in this discussion, Benson made it clear that he thought the three men in the article belonged in a mental hospital.

In April 1960, Gabor announced that he was waiting for a letter to arrive from his wife. Rokeach immediately spotted an opportunity to broaden the experiment, since the wife only existed in Gabor's imagination: he had never been married. Rokeach wanted to find out whether he really believed in her existence, and if so, whether he would renounce his false identity if she asked him to. And so he began writing Gabor letters that he signed 'sincerely Madame Dr R.I. Dung'.

Gabor really was convinced that he had a wife. He dutifully went to the meeting-places mentioned in the letters, where of course she never showed up. About a week after the first letter, he explained to Rokeach that his wife was actually God. Rokeach, alias Madame Dr R.I. Dung, also issued instructions in the letters he sent, telling Gabor, for example, to sing a particular song or share his money with the other men. At the beginning, he dutifully followed his wife's orders, though he never complied with his wife's request to drop the name Dr R.I. Dung.

On 15 August 1961, two years after their first meeting, the three Christs of Ypsilanti – which incidentally was also the title of Rokeach's book on the experiment – met for the last time. Rokeach had abandoned all hope of returning them to normality through therapy. He had also recognized that the three men preferred simply to live in peace with one another rather than trying to resolve the matter of their identities once and for all.

1961 Obedient to the Last

When Morris Braverman walked into the Linsly-Chittenden Hall at Yale University in New Haven, Connecticut in the summer of 1961 he can have had no inkling that, within an hour, he would have been responsible for torturing a person for no reason whatsoever. Braverman, a 39-year-old social worker, had answered an ad in the local paper that ran: 'We will pay five hundred New Haven men to help us complete a scientific study of memory and learning.' The pay 'for approximately one hour's time' was four dollars, plus 50 cents travel allowance. Braverman sent off the application form to the address given in the advertisement. A few days later, he received a phone call inviting him to take part.

What ensued was one of the most controversial experiments in social psychology. Some consider it the most important experiment ever to be conducted on human behaviour, while others think that it should never have been allowed to go ahead. Before long, it came to be known simply as the 'Milgram

Experiment' after the 27-year-old assistant professor, Stanley Milgram (see also pages 180 and 187) who devised it. Nowadays, it is so well known that it is regularly cited in newspaper reports on, say, the genocide in Rwanda or instances of torture in Iraq. In France there is a punk band named 'Milgram', while a New York comedy duo go by the name of 'The Stanley Milgram Experiment'. Stanley Milgram, then, achieved worldwide fame with his experiment, and it cost him his career.

When Braverman entered the laboratory, he was greeted by the person conducting the experiment – a young man in a grey labcoat – and introduced to the second participant, who had arrived before him: James McDonough, a 47-year-old bookkeeper from West Haven. The chief tester began by explaining the purpose of the experiment to the participants: it was designed, he said, to measure the effects of punishment on learning success. To achieve this, one of them would have to play the teacher and the other the student. And so he got Braverman and McDonough to draw lots to determine which role each would play. But what Braverman didn't know was that the draw was rigged: both slips of paper had 'teacher' written on them. In fact, McDonough was an actor who was just pretending to be the second test participant. For the experiment that Milgram wanted

The psychologist Stanley Milgram with the 'electric shock generator', a dummy apparatus built to fool the test subjects into thinking that they were delivering real shocks.

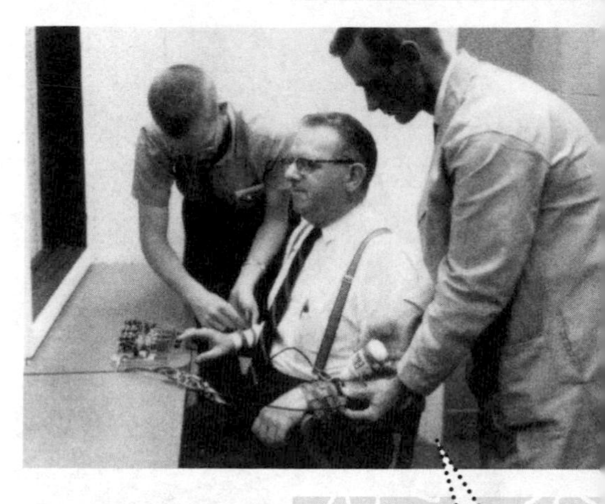

Electrodes being attached to the body of the 'victim', who supposedly had a heart condition – in actual fact, an actor.

to do, it was essential that the unwitting participant – namely, Braverman – play the teacher.

After they had drawn lots, the chief tester took McDonough into a side room, where he strapped him into a chair that bore a passing resemblance to an electric chair. To his left wrist, he attached an electrode, which, he explained to Braverman, was connected to a generator in the main room. McDonough was given just enough freedom of movement with his right hand to allow his fingers to reach a device on the table next to him that had four buttons. When McDonough asked the tester how strong the electric shocks would be, he was told that although they would be 'extremely painful', they would nevertheless cause 'no permanent tissue damage.'

Back in the main room, the tester explained to Braverman what was required of him. Using an intercom system, he was to read out pairs of words to McDonough in the adjoining room, such as 'blue–box', 'nice–day', 'wild–duck' and so on. Then, on a second run-through, Braverman would just give the first word in each pair. McDonough's task was to remember the second word in each case. So, when Braverman said 'blue' and then gave McDonough four choices – 'day', 'box', 'sky' and 'duck' – he had to choose the correct one by pressing one of the buttons.

If McDonough hit the right button, Braverman was to continue with the next word on the list. But if he got it wrong, Braverman was to punish him by giving him an electric shock. The first mistake earned him a shock of 15 volts, the second 30 volts, the third 45 volts, and so on, up to a maximum current of 450 volts. To deliver the shocks, Braverman had a device in front of him that was equipped with a whole row of switches and a manufacturer's serial plate that read 'Shock Generator, Type ZLB, Dyson Instrument Company, Waltham, Mass., Output 15 Volts – 450 Volts. If Braverman had known his way around Waltham, Massachusetts he would have known that there was no such firm of that name there.

madsciencebook.com **The original footage of Milgram's experiment is one of the most striking documents in the whole of social psychology.**

Milgram got the idea for this experiment in 1960, when he was still a student at Princeton University in New Jersey. His tutor

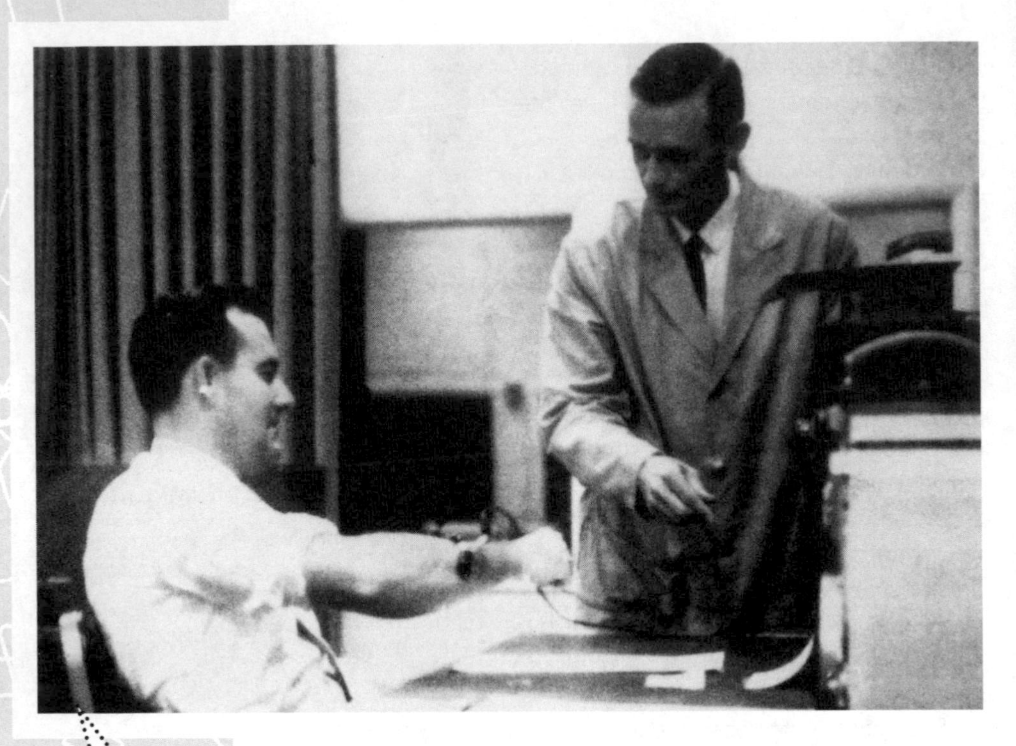

The test subject being ordered to deliver electric shocks to the victim in an adjoining room, as punishment for getting answers wrong.

there, the psychologist Solomon Asch, demonstrated through another experiment that later became extremely well-known the enormous pressure that a group can exert on an individual. The participants in Asch's experiment gave answers that they knew must be wrong in a guessing game just so as to fall in line with the rest of the group.

Milgram was keen to follow this experiment up by testing the effect of group pressure in a less innocuous situation. Could a test participant be induced to cause pain to another person for no reason whatever? Milgram conducted some preliminary tests to find out how far participants would go in the absence of group pressure. What emerged from these tests was that there was absolutely no need for a group: a single person would do.

Braverman knew nothing of all this when he gave McDonough a 15-volt shock after his first mistake. McDonough made more mistakes and Braverman steadily cranked up the voltage, 15 volts at a time, just as he had been instructed to prior to the experiment.

After the 120-volt shock, McDonough told the evaluator over the intercom that the shocks were now becoming painful. At 150 volts, he shouted 'Experimenter, get me out of here! I won't be in the experiment any more! I refuse to go on!' At 180 volts: 'I can't stand the pain!' When the current reached 270 volts, McDonough yelled out in pain and said that he wouldn't answer any more questions.

Braverman turned to the test evaluator, who simply said 'Please go on,' instructing him to treat the lack of an answer as though it were a wrong one and punish the student with a shock just the same. Braverman shifted about nervously in his chair and gave a couple of wheezy laughs, but went on nevertheless. McDonough now refused to give any more answers, but just shrieked with every electric shock.

Braverman turned to the tester once more and asked him: 'Do I have to follow these instructions literally?' The latter replied 'The experiment requires that you continue.' And so Braverman continued. After 330 volts, McDonough fell silent. Braverman made a half-hearted offer to swap places with him, before continuing with the shocks. Beneath the rocker switch for delivering 375 volts was a warning sign that read 'Danger: Severe Shock'. Yet Braverman went on, right up to the final rocker switch delivering 450 volts.

The Milgram Experiment even found its way into pop culture. Shown here is the cover of the French–German punk band Milgram's album *Vierhundertfünfzig Volt* [450 Volts] (an allusion to the maximum electric shock delivered in the test).

New Haven social worker Morris Braverman wasn't the only person who delivered potentially fatal electric shocks in the summer of 1961, just because some experimenter who didn't wield much authority ordered him to do so. Other test subjects, such as the manual labourer Jack Washington, the welder Bruno Batta, the nurse Karen Dontz and the housewife Elinor Rosenblum also went right to the end of the scale. A total of more than 1000 test subjects took part in Milgram's experiment in its various guises. Two-thirds of them went as far as delivering a 450-volt shock.

Milgram was wrong-footed by this result; indeed, everyone was. In subsequent lectures, he described the bases of the experiment in detail and then asked his audience what they thought the outcome would be. Neither psychologists nor laymen came even close to predicting quite how ready people were to follow orders blindly. Most reckoned that nobody would go beyond 150 volts.

Milgram knew that his experiment was sensational stuff, but from a scientific point of view there was a problem with it: it neither solved a problem nor confirmed a theory. Psychology journals turned it down for publication twice. It was only when Milgram, at a third attempt, outlined various different versions of the experiment and compared them with one another that it finally appeared in print, under the title 'Behavioral Study of Obedience' in the *Journal of Abnormal and Social Psychology* in 1963.

Milgram conducted the experiment in almost 20 different variations. In one variant, the student complained that he had a weak heart, while in another the experiment took place in a run-down office block off the university campus. Sometimes, women would deliver the electric shocks. But however it was tweaked, the outcome remained the same: over half the test subjects delivered the maximum shock.

In other versions of the experiment, the student was in the same room as the 'teacher' test subject. Although in this instance the degree of compliance did decline noticeably, it was remarkable that even when the experimenter ordered the test subject to personally press the student's hand down on the metal plate that transmitted the current, one-third of them still went right up to the 450-volt mark. Certainly the physical proximity to the victim seemed to play an important role, but a more significant factor was the presence of the evaluator. When he switched to giving his orders over the phone, only one in five subjects complied.

No sooner had Milgram published his experiments than they became public knowledge the world over. Newspapers picked up on the story and placed their own interpretations on the results. The burning question was whether people in real life acted in the same way as the browbeaten test subjects. There is still no consensus on this. Milgram himself constantly related the experiment to Nazi atrocities in the Second World War. Since the end of that conflict, the world had been looking for ways to explain the Holocaust. Milgram was convinced that one possible explanation was the propensity to follow orders inherent in all of us.

Publication of his study coincided with reports by the philosopher Hannah Arendt on the recent trial in Jerusalem of the Nazi war criminal Adolf Eichmann. In the famous articles she wrote for the *New Yorker* magazine, Arendt introduced the concept of the 'banality of evil'. She claimed that Eichmann wasn't the sadistic monster that the prosecution tried to paint him as, but rather an unimaginative bureaucrat who had simply been doing his duty.

This chimed in perfectly with the findings from Milgram's experiment. His test participants were neither especially aggressive nor did they derive any pleasure from delivering electric shocks to the students. Quite the opposite, in fact: many of them were anxious, began to sweat, or argued with the experimenter; however, only very few were strong-willed enough to curtail the experiment. It was clear that people viewed disobedience as such a radical act that they would rather jettison their fundamental moral convictions. Milgram concluded that 'the key of the behavior of subjects lies not in pent-up anger or aggression but in the nature of their relationship to authority.'

In September 1961, shortly after the shocking results began to emerge, Milgram wrote a letter to his funding body, the National Science Foundation, in which he observed: 'I once wondered whether in all of the United States a vicious government could find enough moral imbeciles to meet the personnel requirements of a national system of death camps, of the sort that were maintained in Germany. I am now beginning to think that the full complement could be recruited in New Haven.'

Milgram's linking of the experiment to the Holocaust made him a controversial figure, yet far more damaging was the criticism that his experiment had been unethical – specifically, the question of how much a stress a test participant should be subjected to. Some of his colleagues thought that he had gone too far. While Milgram had anticipated such criticisms being levelled at him, he was nevertheless disappointed that no credit was given to the care with which he had designed the experiment.

At the end of the hour-long experiment, the student was brought in from the next room and the test subject was told that he hadn't in fact received any electric shocks. In a follow-up study, Milgram finally quizzed all those who had taken part about their attitude to their involvement. Fewer than 2 per cent said that they wished they'd never got involved. Even so, there would be no possibility of repeating the experiment nowadays: the furore that erupted over Milgram's test saw stringent ethical guidelines on the admissibility of experiments imposed at all universities.

madsciencebook.com

The psychologist Thomas Blass published the first biography on Milgram's remarkable life: *The Man Who Shook the World: The Life and Legacy of Stanley Milgram* Basic Books, 2004.

Source

Milgram, S. (1974), *Obedience to Authority. An Experimental View.* Harper & Row.

Milgram, S. (1963), *Behavioral Study of Obedience. Journal of Abnormal and Social Psychology* 67 (4), pp. 371–78.

Very few of those who were directly involved are now willing or able to explain what happened during the experiment. Out of the 1000-plus participants, those who are still alive are reluctant to talk about it. Milgram's data, rendered suitably anonymous, is now salted away in card-index files in the library at Yale. All the names of participants that crop up in connection with the experiments have been changed – even in this article.

One of the few contemporary witnesses is Milgram's research assistant Alan Elms, now professor of psychology at the University of California. He says that many people still react with a mixture of fascination and revulsion when they find out that he was involved in the experiment.

Milgram paid a high price for revealing to mankind an unpalatable truth about itself: he was later appointed assistant professor at Harvard University but could never secure a permanent post there. In 1967, he transferred to the far less prestigious City University of New York, where he died of heart failure in 1984, aged 51. Shortly before his death, his wife became a grandmother for the first time. She told a reporter that her grandson's second name was Stanley. When the reporter asked her why that wasn't the boy's first name, she replied: 'I think it would be a burden to go through life with the name Stanley Milgram.'

1962 High on Good Friday

The Good Friday service in Easter 1962 was a memorable experience for ten seminarians at the Andover Newton Theological School. Although they could remember hardly anything of the sermon delivered by Pastor Howard Thurman, they could recall a sea of colours, voices from the Beyond, and the feeling that they were melting into the surrounding world. In a word, the students were high.

At the beginning of the 1960s, some daring scientists turned their attention to studying mind-altering substances. This was the period when it was all part and parcel of a lecture on mysticism to ingest magic mushrooms to gain practical insight into the subject, and when a doctoral thesis could entail giving students drugs and observing their behaviour. This is exactly what Walter Pahnke did: this young theologian and doctor from Harvard University was keen to discover whether psychedelic drugs could induce the kind of mystical sensations that only very few people otherwise experience, for example when in a state of religious trance. Users of LSD, psilocybin or mescaline had long claimed that this was the case.

Pahnke turned to Timothy Leary, who a short time earlier had begun conducting drug experiments at Harvard, and who later became a leading figure in the 1960s counterculture. He proposed an experiment to Leary: test subjects would attend a church service, but half of them would be given mind-expanding drugs in advance. Afterwards, all participants would be required to fill in a questionnaire and be interviewed. Comparing the findings with descriptions of mystical experiences from the realm of religion would demonstrate whether there was a qualitative difference between them.

Leary was half shocked and half amused by the idea. As he later wrote in his autobiography: 'If he had proposed giving aphrodisiacs to twenty virgins to produce a mass orgasm, it wouldn't have sounded further out.' He explained to Pahnke that a psychedelic trip was an intensely personal experience and that a person would have to have experienced several himself before he could even contemplate devising such an experiment. However, Pahnke was adamant that he would have to wait until his thesis had been accepted before he indulged. He didn't want anyone accusing him of partiality: the experiment would have a chance of succeeding only if he hadn't taken any drugs himself beforehand.

Leary was impressed by Pahnke's stubbornness and so agreed to set up a small test at his house with a couple of theological students. In his autobiography, he reported that each of the subjects experienced 'visions as dramatic as Moses or Mohammed' and that 'it was strong Old Testament stuff.' One of them was afraid he was going to die, while another spent the trip 'copulating the carpet'. None of this fazed Leary: 'So there were crises of conscience and identity – but it was all healthy and yeasty.'

After Pahnke and Leary had established an exact procedure, the main experiment could then take place. On the morning of Good Friday, two hours before the service, 20 students met in the crypt of Boston University's Marsh Chapel. They were encouraged 'not to try to fight the effects of the drug even if the experience became very unusual or frightening.'

In groups of four, they were then asked to wait in separate rooms to be given the capsules containing psilocybin – magic mushroom in powdered form, a drug that was also used by certain indigenous peoples during their sacred rites. Each group had two minders. The night before, someone not involved in the experiment had packed the capsules – for each group, two containing the drug and two placebos. Pahnke was concerned to conduct his experiment along the most stringent lines governing a clinical drug trial – namely double-blind. In other words, to ensure that the data that emerged could be evaluated totally impartially, neither the evaluators nor the subjects would know who had actually taken the magic mushrooms. He even employed a further smokescreen: the placebo capsules didn't contain the customary ineffective powder but 200 milligrams of niacin, a vitamin that induces hot flushes and that would thereby simulate for the participants the effect of having taken the psilocybin. But it soon became apparent quite how pointless it was to try and conduct an experiment with psychedelic drugs double-blind. Although the effect of the niacin was to produce some initial confusion, it pretty soon became clear who was in which group. In other words, the list given to the evaluator to show who had taken the placebo and who was on the real drug was rendered totally redundant.

Before Timothy Leary became a leading light of the 1960s counterculture, he carried out drug experiments at Harvard University.

The five groups were then taken to attend the service in the small crypt chapel, where Father Thurman's voice boomed out through a loudspeaker. He was delivering his regular Good Friday sermon in the main chapel one floor above. Ten of the twenty test subjects sat attentively on their pews. Of the other ten, some wandered about the chapel muttering to themselves, one lay on the floor while another sprawled across a pew, and another sat down at the organ and played dissonant chords. Five of the ten minders also started behaving oddly. Overruling Pahnke's misgivings, Leary had also insisted that they be given the drug. He justified his decision by claiming, 'We are all in it together. Shared ignorance. Shared hopes. Shared risks.'

The service lasted two and a half hours. When it had ended, the students were interviewed for the first time. At 5 o'clock, Leary invited everyone to come and eat with him, but 'the trippers were still too high to do much except shake their heads, saying "Wow!"', as he later recalled.

Source

Doblin, R. (1991) "Pahnke's 'Good Friday Experiment': A Long-Term Follow-Up and Methodological Critique." *The Journal of Transpersonal Psychology* 23 (1), pp. 1–28.

Pahnke, W. and Richards, W. (1966), 'Implications of LSD and Experimental Mysticism.' *Journal of Religion and Health* 5 (3), pp. 175–208.

In the days following the experiment, and again six months later, the subjects were quizzed about what they had gone through. Pahnke wanted to use his questionnaire to find out how intense a mystical experience they had had. This consisted of questions relating to nine separate realms of experience, including the sensation of being in harmony with oneself, the impression that time and space were being transcended, along with questions relating to people's moods, their sense that things were ineffable and their feelings of transience. The results were unequivocal: eight of the ten students who had eaten the magic mushrooms experienced at least seven of the impressions and feelings customarily associated with a mystical experience. By contrast, no one from the control group reached this kind of score. In every category, they lagged far behind the experimental group.

The difference also manifested itself clearly during the interviews. The students on psilocybin claimed that the trip had also had a positive knock-on effect on their daily lives: they maintained that it had raised their consciousness, caused them to reflect more deeply on their attitude to life and made them more socially aware. Pahnke reckoned that these positive effects could be attributed to the fact that the church service had given the participants a familiar framework in which to put their drug experience.

So, it seemed that ingesting 30 milligrams of white powder could bring about a state of consciousness identical to that experienced by Christians, Buddhists or Hindus after self-castigation, withdrawal from the world to live as a hermit and long years of meditation. This was indeed a bold claim: 'To some theologians, the awareness that it appears possible to experience mystical consciousness [samadhi in Advaitan Hinduism, satori in

Zen Buddhism, the beatific vision in Christianity] with the help of a drug on a free Saturday afternoon at first appears ironic and even profane,' Pahnke noted. However, for him this possibility indicated only 'the poor methods that have been used by men to gain such experience.'

Pahnke was well aware that psychedelic drugs were a highly emotive issue in the Church. The experiment not only raised the questions of whether mystical experiences were solely based on neurological processes and whether divine inspiration was really just physical brain chemistry. It also cast doubt on the principle that a mystical experience had to be earned through asceticism. Even so, Pahnke firmly believed that research into these new states of consciousness had a great future. He dreamt of an institute where psychologists, psychiatrists and theologians would be able to conduct experiments to fathom the mysteries of mysticism. But things didn't quite pan out that way: although Pahnke's doctoral thesis was accepted, funding for further experiments dried up. Psychedelic drugs were banned, since the public health authorities regarded them as dangerous. Leary was fired. As for Pahnke, he was killed in a diving accident in 1971.

Twenty-five years after the experiment, the psychologist Rick Doblin attempted to find the surviving participants. In four years of detective work, he succeeded in tracking down 19 of the 20 students. Sixteen of them agreed to be interviewed and filled in the same questionnaire as in the original experiment. The results were astonishingly consistent: those in the experimental group and the control group gave much the same answers as they had done a quarter of a century before. The test subjects from the experimental group described the Good Friday service of 1962 as one of the high points in their spiritual lives. They all claimed that the experiment had had a positive influence on them. Some attributed their later socially aware outlook to it, while others said it had helped them come to a positive accommodation with their fear of death.

Nevertheless, most of the former participants also recalled that the experiment also had its negative aspects. There were moments when they thought they were going mad or dying. Pahnke treated only this aspect in passing in his thesis. In particular he hushed up the fact that one subject had to be injected with an antidote when the situation got out of hand: seized with an urge to put Pastor Thurman's call to spread the word of Christ into action straight away, one student left the chapel and went out onto the street, from where he had to be fetched back.

Despite this, Doblin's assessment of the experiment is largely positive. Though the men from the experimental group weren't in favour of a complete liberalization of drugs, they did take the view that drugtaking in the right context could be a thoroughly enriching experience.

Just one member of the control group claimed that the experiment had benefited him greatly. Not that it was the church service as such that had such a positive effect on him, but rather the decision he made during it to try psychedelic drugs himself at the next available opportunity.

1962 Cookie-cutters Provide New Insights

At first glance, the experiment by the American psychologist James J. Gibson wasn't exactly destined for lasting fame in the annals of science. It was designed to corroborate something that had long been known about and was so simple that anyone could reproduce it instantly with a few cookie-cutters from the kitchen drawer. Yet Gibson recognized the deeper significance behind the obvious result, and in the process brought about a paradigm shift in research into human perception.

The experiment went as follows. Subjects were required to put their hands under a cloth and pick out a specific shape of cookie-cutter from among six differently shaped such items. The success rate was 95 per cent when they were allowed to pick up the cutter and feel it, but if the cutter was pressed into their upturned palm without them holding it themselves, then the strike rate fell to just 49 per cent. On the face of it, no great surprise. Who could dispute the fact that the most reliable and quickest method of recognizing a particular shape is to feel it with your fingers?

Yet Gibson realized quite how strange this result actually was. Say the shape in question was a star. The brain's task should by rights have been far simpler if the star was pressed down onto a person's skin without them moving. In this case, the image that the receptors on the skin conveyed to the brain would correspond exactly to the representation of a star. By contrast, if the person's fingers actively felt the star, their fingertips would swamp the brain

Source

Gibson, J.J. (1962) Observations on Active Touch. Psychological Review 69 (6), pp. 477–491.

with a confused mass of successive neural signals that would have nothing to do with the shape of a star and which, if the person felt the star for a second time, wouldn't be the same twice over. Yet despite this, active touching yielded double the success rate.

Could this really be correct? Perhaps, Gibson speculated, the discrepancy in the success rate came about only because the fingertips are more sensitive than the palm, and so he devised another experiment. In this, the subjects had the cookie-cutters pressed into their palms just as in the first experiment. But this time the cutters were either held still or were gently but rapidly rotated around their own axis, a short distance to both the left and the right. In both cases receptors in the same patch of skin were active and were just as sensitive on each occasion. Just as when the objects were felt, when they were rotated, the success rate went up from 49 to 72 per cent.

Gibson summed up the paradoxical situation thus: 'The form of the object seemed to become clear when the form of the skin-deformation was most unclear … a clear unchanging perception arises when the flow of sense impressions changes most.'

Gibson could find only one explanation for this state of affairs: the prevailing notion of how tactual perception functions was wrong. In essence, then, our sense of touch is not merely the passive transmission of tactile stimuli but rather the active exploration of shapes, which produces a whole stream of shifting stimuli. The brain is evidently quite capable of filtering out from these constantly changing sensory impressions those permanent structures that correspond to the physical make-up of the world around us.

1963 The Lost Letters

Imagine you're strolling around your neighbourhood and you come across an unposted letter with a stamp on it lying on the road, addressed to the 'Friends of the Nazi Party'. Would you post it? What if it was addressed to the 'Friends of the Communist Party', the 'Medical Research Association' or simply to one Walter Carnap?

Several inhabitants of the small town of New Haven, Connecticut were confronted with precisely these questions in the spring of 1963, when they found letters like this lying around. But what these passers-by didn't know was that the letters hadn't, as seemed to be the case, simply been dropped by accident. Rather, students from Yale had placed them quite deliberately around the town; on the street, in telephone boxes, in shops and under cars' windscreen wipers (with a handwritten note in pencil reading 'found near your car'). They distributed the letters in such a way as to make it very unlikely that a single person would find two of them. For if they did, they'd have noticed that, although the addressees' names were different, the address itself was always the same: 'P.O. Box 7147, 304 Columbus Avenue, New Haven 11, Connecticut'.

Two of the 'lost' letters that Stanley Milgram spread around. He used them as an unobtrusive way of probing people's attitudes to various topics.

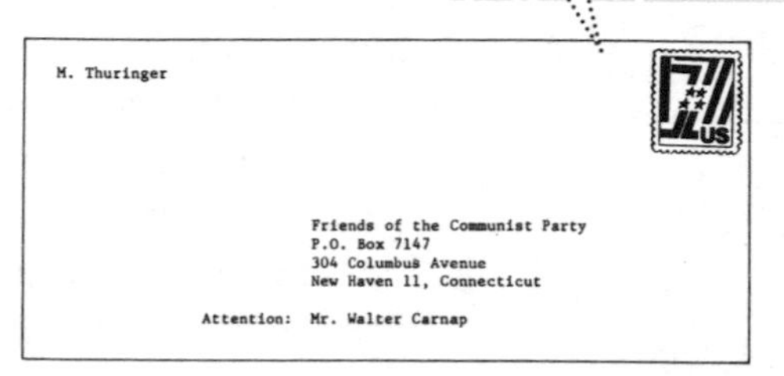
M. Thuringer

Friends of the Communist Party
P.O. Box 7147
304 Columbus Avenue
New Haven 11, Connecticut

Attention: Mr. Walter Carnap

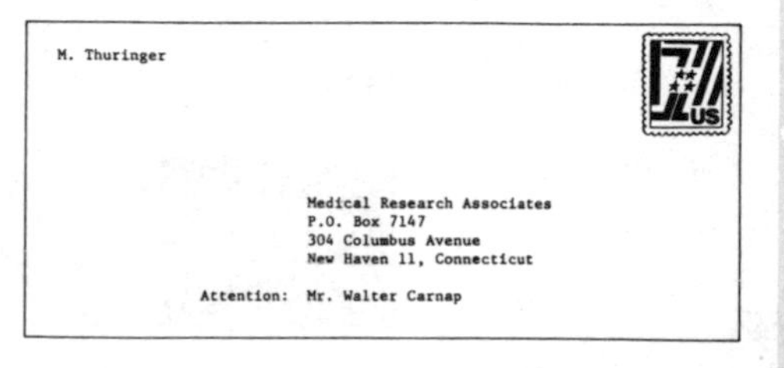
M. Thuringer

Medical Research Associates
P.O. Box 7147
304 Columbus Avenue
New Haven 11, Connecticut

Attention: Mr. Walter Carnap

P.O. Box 7147 had been rented out by the psychologist Stanley Milgram (see pages 168 and 187). A fortnight after distributing the envelopes, out of batches of 100 letters to each of the four addressees, 25 had been delivered to the National Socialist Party, likewise 25 to the Communist Party, while 72 had made it through to the Medical Research Association and 71 to Walter Carnap.

Milgram was pleased with this result. The difference in the number of returns indicated that the 'lost-letter-technique' could be used to collate people's attitudes to particular organizations and hence to particular topics.

Traditional studies in this field had taken the approach of quizzing people directly or asking them to fill out questionnaires. The problem with this was that you could never be sure whether they were telling the truth. Especially when it came to contentious subjects, the results of surveys seldom reflected the real state of affairs. This wasn't the case, however, with Milgram's letters: because people had no idea they were taking part in an experiment, they made no attempt to disguise their views.

At first sight, Milgram wrote, this method might appear to be just a dodge for lazy social psychologists, given that it consisted of little more than leaving a few letters lying around and then waiting for them to be returned. In actual fact, though, distributing

Source

Milgram, S. et al. (1965), 'The Lost-Letter-Technique: A Tool of Social Research.' *Public Opinion Quarterly* 29, pp. 437–438.

hundreds of letters was an extremely arduous business. In order to ensure that the experiment yielded reliable data, each and every one of the letters had to be positioned individually by hand.

Milgram did try to streamline the procedure. On one occasion, he drove around at night throwing them out of the window of his car, but they often came to rest face-down. Another time, he scattered them from a light plane over Worcester, Massachusetts, but to little avail: many of the letters landed on house roofs, or in trees or ponds. Not only this – they also got caught in the ailerons of the Piper Colt aircraft, 'endangering not only the results but the safety of the plane, pilot and distributor.'

Thereafter, the lost-letter technique was employed in hundreds of experiments. Milgram himself used it successfully to predict the winner of the 1964 US presidential race, Lyndon B. Johnson, even though he greatly underestimated Johnson's share of the vote. The technique is particularly well suited to extremely controversial topics. For example, it has recently been used to canvass opinions on creationism, sex education and gay teachers.

1964 Bullfighting by Remote Control

madsciencebook.com Original footage of Delgado's bullfight.

The location was well chosen: where should a Spanish neuroscientist demonstrate his control over an animal's brain functions if not at a bullfight? And so it was that, on a spring evening in 1963, José M.R. Delgado came face to face with Lucero, a 250-kilogram fighting bull owned by the landowner Ramón Sánchez, who had granted Delgado the use of a small practice ring on his estate of La Almarilla in Córdoba for the experiment. To begin with, several experienced bullfighters psyched up the animal, while Delgado kept his distance behind a wooden hoarding. But then it was the professor's turn. Although his experience with the muleta (the red cloth used by bullfighters), which he had gained during his youth at various village fêtes, was 'pretty limited', as Delgado himself later wrote, he was adamant that a researcher must take responsibility for his chosen methods and be prepared to face the bull himself. And so, wearing a shirt and tie, he appeared from behind the hoarding, walked tentatively up to Lucero, waving the red cloth in his right hand to provoke the bull, and holding a remote-control device in his left. A few days before the experiment, he had placed radio-controlled electrodes in the brain of Lucero and several other bulls. Now, as Lucero lumbered towards him, Delgado suddenly dropped the cloth and pressed a button on the remote-control. The brain implant activated, sending a milliampere of alternating current through Lucero's brain. This instantly dissipated the animal's aggression; Lucero skidded to a halt and trotted off at a leisurely pace.

The neurologist José Delgado stops a bull in its tracks by using a remote control to activate electrodes implanted in its brain.

Afterwards, the only concern among Spanish newspapers was that the experiment might herald the end of the Corrida (bullfight). Under such headlines as 'Remote-control Bullfighting' and 'They're Going to Take Our Toreros [Bullfighters] Away', they reported the experiment in distinctly unflattering terms. By chance, two years later, Delgado's experiment also made it onto the pages of the *New York Times*; the professor was giving a lecture in New York and a journalist from the newspaper was in the audience. This report instantly caused ripples. As Delgado later said, 'Since that time, I've received mail each year from people who think I'm controlling their thoughts.'

As a young research scientist, Delgado had emigrated from Spain to the United States, and at the time of the experiment was a professor at Yale University. He wanted to find out more about human and animal behaviour by stimulating the brain with electrical pulses. As he had done with many of the animals he used in his research, he implanted electrodes into the bull's brain in order to try and induce certain modes of behaviour. For example, he had already succeeded in making monkeys yawn and cats attack just by pressing a button. He had also found that he could influence affability, speech fluency and anxiety in epilepsy sufferers.

Delgado was not only convinced that electrical stimulation of the brain was the key to understanding the biological bases of social behaviour, but was also the prophet of a new 'psychocivilized' society, whose members would have at their fingertips a technology that would allow them to become 'happier, less destructive and better balanced men.'

"I've received mail each year from people who think I'm controlling their thoughts"

Colleagues called Delgado either a 'mad scientist' or 'the Thomas Edison of the brain'. He countered critics who feared total control over human beings by citing the old adage that knowledge in itself isn't bad, just its application. He asked doubters to consider the following scenarios: 'Suppose that the onset of epileptic attacks could be recognized by the computer and avoided by feedback: would that threaten identity? Or if you think of patients displaying assaultive behaviour due to abnormalities in brain functioning: do we preserve their individual integrity by keeping them locked up in wards for the criminally insane?'

While Delgado's dream of a psychocivilized society hasn't yet become a reality, electrical stimulation of the brain certainly is, after a long period of neglect, being used on people: for instance, in helping patients with a variety of neurological conditions such as Parkinson's disease manage their symptoms. In the summer of 2007 it also emerged that a man who had lain in a coma for six years had been brought back to consciousness by electrical brain stimulation.

Source

Delgado, J M R. (1982). 'Toros Radiodirigidos', J M De Cossío and A. Díaz-Cañabate in *Los Toros*. Madrid, Espasa-Calpe. pp. 186–210.

1966 Psychology with Car Horns

Students Alan E. Gross and Anthony N. Doob didn't really have much of an idea what they wanted to study. All they knew was that they had to conduct an experiment. This was a course requirement in the Social Psychology Faculty at Stanford University, which they both attended. The only thing stipulated in advance was the experimental method.

Not long before, a book had appeared pointing to the deceptiveness of research in the laboratory: test subjects behave differently when they know they're being observed. Questionnaires give them ideas that they would never have arrived at if they hadn't been posed leading questions in the first place. Therefore, it was the task of the Stanford students to conduct an unobtrusive experiment, a test in a natural setting that the subjects wouldn't even notice they were participating in.

Gross and Doob thought long and hard about the easiest place in which to carry out a natural experiment but could only come up with ideas that either would have involved expensive equipment or violated the privacy of the participants. One afternoon, when they were discussing frustration and aggression, they suddenly realized where these feelings most commonly manifested themselves – in a traffic jam.

That same day, the pair drove around Palo Alto in Doob's 17-year-old Plymouth and remained stationary several times when the light changed to green. The drivers behind them reacted instantly, providing the students with an easily measurable yardstick for their frustration, namely the time it took for them to sound their horns.

However, this didn't yet constitute an experiment. After all, in an experiment, you usually want to ascertain how its outcome – i.e. the effect it produces – differs when it is conducted under various conditions. Gross and Doob had already discovered the effect (also known as a 'dependent variable'): namely, frustration as measured by the time it took for a person to sound their horn. But what was to form the independent variable in their experiment? In other words, how might they vary the conditions and then measure the resulting degree of frustration? Their first thought was the number of occupants in the car blocked by the stationary vehicle, but they had no way of influencing this. Then there was the sex of the driver of the car blocking the road, but none of their female fellow students wanted to risk taking part in the experiment. So finally they lighted on the status of their car: would the behaviour of the driver behind them differ if he found himself blocked not by a cheap student's car but by an expensive, high-status car?

This immediately led to the next problem: where to get hold of an expensive car. A fellow student owned a new black Cadillac Fleetwood, but he had no intention of swapping it for the day for a rusty old 1949 Plymouth. Gross and Doob ended up renting a brand-new Chrysler Crown Imperial Hardtop from Avis. For their low-status cars, the used a rusty Ford Caravan and a grey Rambler Sedan. Since neither of them knew much about cars, they also solicited the help of a couple of high-school students, who could tell them, in each case, the make and model of the car they stopped in front of.

By 20 February 1966 everything was in place. From 10.30 a.m. to 5.30 p.m. Gross and Doob took turns in using one of their three cars to snarl up six intersections in Palo Alto and Menlo Park. Meanwhile, lying down on the back seat, the one who wasn't driving measured the time between the lights changing to green and the first horn, and then the time between the first and second horns. If, by that time, the lights were still green, the driver would move off.

In the process, Gross and Doob learned that unobtrusive experiments aren't without their dangers: two drivers of blocked cars didn't even bother honking but instead shunted into the back of the experiment car. In these cases, Gross and Doob didn't feel like hanging around to wait for the honking to start.

The results were clear: men who found the rusty Ford in front of them sounded their horn after an average of 6.8 seconds, whereas it took 8.5 seconds with the Chrysler Crown Imperial. Women exhibited the same pattern of behaviour, but were generally more restrained. A further analysis showed that the old Ford was honked twice by eighteen drivers, while the new Chrysler was honked twice by only seven drivers.

Several journals turned down the experiment for publication. It was the chief editor of the *Journal of Social Psychology* who finally recognized that it was an 'ingenious idea for research'. He was to be vindicated in his judgement: Gross and Doob's work was subsequently published in many textbooks and has spawned a veritable glut of car-honking studies. For example, researchers

Source

Doob, A.N. und Gross, A.E. (1968), 'Status of Frustrator as an Inhibitor of Horn-Honking Responses.' *Journal of Social Psychology* 76, pp. 213–218.

investigated whether women were honked more than men (in the USA, the answer was yes, but in Australia, no), or what happened when cyclists held up cars or when the offending vehicle was a pick-up truck with a rifle clearly visible in the back window. An additional finding was that it took longer for the first horn to sound if a scantily-clad woman was standing by the side of the road (see p. 239).

1966 Tip for Hitchhikers 1: Get an Injury!

The first experiment on the attitude of car-drivers towards hitchhikers is thought to be that which is described in the work 'Helping and Hitchhiking' by one James H. Bryan from Northwestern University. A 'male student, clean-shaven, with short-cropped blond hair and dressed in shorts, a white T-shirt and sneakers' (probably a description of Bryan himself) stood beside a four-lane highway in Los Angeles on four summer days and tried to get a lift. Sometimes he had a bandaged knee and a stick, and sometimes not.

The results yielded the first bit of scientific advice for hitchhikers: wear a bandage and carry a stick on principle! In this way, Bryan was invariably able to at least double his chances of being offered a lift. However, the next hot tip that research came up with for hitchhikers is not, unfortunately, available to all and sundry, at least not without surgery (see p. 223).

Source

Bryan, J.H. (1966) "Helping and Hitchhiking.' Unpublished manuscript.

1967 Six Steps to Knowing Everyone

The problem had long been common currency among mathematicians: pick two people at random from any two places you like in the world. How many friends, friends of friends, and friends of friends of friends does it take to establish a connection between them? In short: how closely are any two people on Earth connected? How small is the world?

At first sight, the solution to this 'small-world-problem', as it is also known, is quite simple: if you know how many people on average a particular individual knows, then you can do a straightforward projection. Say, for example, I know ten people, and each of them in turn knows ten others, then in just two steps I am already connected with 10 times 10 people, in other words 100 people. In three steps, I'm connected with 1000, in four with 10,000 and so on.

However, the two mathematicians who did this calculation in the 1950s – Ithiel de Solla Pool from the Massachusetts Institute of Technology (MIT) and Manfred Kochen from the computer firm IBM – encountered two fundamental problems. One of them appeared quite solvable: there were no statistics available on the average number of people known to an individual. Accordingly, several people were engaged to keep a record of their contacts over a period of 100 days. On average, each person amassed 500 acquaintances during that time. But the second problem seemed really insuperable: it's highly likely that several friends of my friends know one another directly. As a result of these mutual friends, then, in the example above, I don't actually acquire ten times more acquaintances with every step, but significantly fewer. How many fewer depends on how closed the circle is in which I and my friends, and in turn their friends, move and also upon how these groups interconnect. With an average of 500 acquaintances, the matter becomes so complicated after just a few stages that de Solla Pool and Kochen decided not to publish the paper that they wrote on the subject in 1958. As they later wrote, 'We never felt we had "broken the back of the problem".' Yet their provisional findings did indicate that people are connected with one another by very few steps.

When the psychologist Stanley Milgram (see pp. 168 and 180) heard about this result, he set about trying to verify it. Milgram's experiment later became so popular that it developed into a party game, while a play was also named after the result.

The experiment found its way into popular culture through the play *Six Degrees of Separation* (1990) by John Guare.

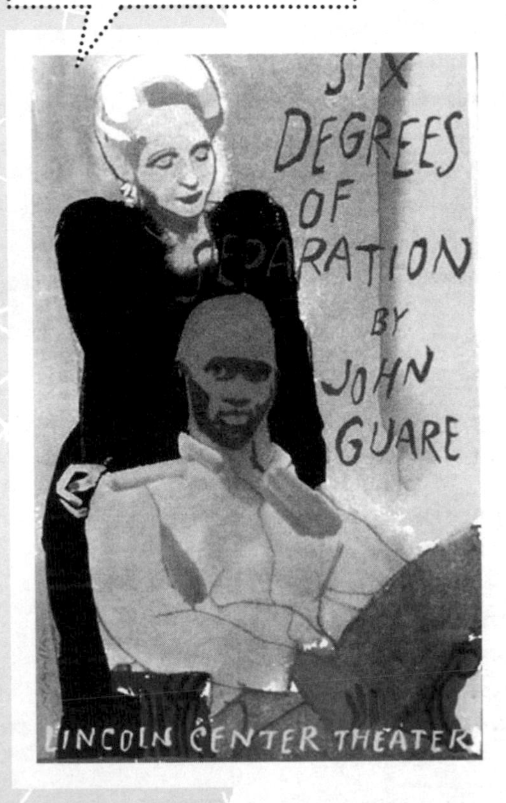

Milgram began by selecting an appropriate target: the wife of a theology student in Cambridge, Massachusetts, where Milgram was currently employed, at Harvard University. He then took as his starting points a couple of dozen people from either Wichita, Kansas or Omaha, Nebraska. They were given the name of the target person, together with a brief description of her and an introduction to the purpose of the study: 'if you do not know the target person on a personal basis, do not try to contact him directly. Instead mail this folder ... to a personal acquaintance who is more likely than you to know the target person ... it must be someone you know on a first name basis'.

The first letter reached the target person after four days. It had started its journey with a farmer in Kansas, who had sent it to the pastor in his home town. The pastor sent it to a colleague in Cambridge, who personally knew the wife of the theology student. And so the letter had reached its target in just two stages. This was one of the shortest chains ever observed by Milgram. Remarkably, in his paper Milgram doesn't give any more results from his first study. The average from his second experiment was 5.5 stages.

The insight that the world really is this small found its way into popular culture by a number of extraordinary routes. In 1990 the American author John Guare published his play *Six Degrees of Separation*, which alludes indirectly to Milgram's experiment and was later filmed with Will Smith in the leading role. In 1994, three students from Albright College in Pennsylvania invented the game 'Six Degrees of Kevin Bacon', which involves connecting any movie actor you care to name with Kevin Bacon via as few films in common as possible, For instance, Will Smith played alongside Laurence Fishburne in *Welcome to Hollywood* (2000), while Fishburne himself then co-starred with Kevin Bacon in *Mystic River* (2003). This result, then, gives Will Smith a 'Bacon number' of 2.

The fact that every rice-farmer in China is connected with Madonna by only a very few stages has caught many people's imagination. In the Czech Republic, there is even a heavy-metal band called Six Degrees of Separation. However, despite the fact that mathematics had made great strides in recent times, and people from all walks of life – for example, computer networking experts and epidemiologists – have become interested in the principles of the 'small world', the basic problem still hasn't been cracked.

madsciencebook.com

Play 'Six Degrees of Kevin Bacon' on the web.

Source

Milgram, S. (1967) 'The Small World Problem.' *Psychology Today* 1 (1), pp. 60–67.

So, it still remains unclear whether Milgram's figure of 5.5 links was correct. He did not publish his experiments in the usual way, in a scholarly journal, but in the popular science magazine *Psychology Today*. The data in his article are sketchy and are hardly verifiable. For instance, Milgram cited his success with the Kansas farmer, whose letter made it to the woman in Cambridge in just two stages. But more detailed information on this study can be found only in unpublished archive material: it turns out that only three of the sixty envelopes distributed to the participants in Kansas reached their destination – and even then only via an average of eight stages. Milgram, who died in 1984, hit on the figure of 5.5 in later experiments, when it was often the case that participants were deliberately chosen who had an extensive social network.

In 2003, scientists from Columbia University in New York repeated Milgram's experiments with e-mails rather than letters. As targets, they chose 18 people in 13 countries. As with Milgram, only a small proportion of the chains that they started were ever completed (384 out of 24,163). The average length from the starting person to the target was 4.05 stages. However, this figure is deceptive, since many of the chains never extended to the target person. An approximation procedure can be used to make up for the missing chains, but only if you know at what stage contact was broken off. The final number that emerged from this study was between five and seven stages. This is astonishingly close to Milgram's six stages, but still doesn't constitute an unequivocal confirmation. And, ultimately, the participants in this experiment were far removed from the average among the world's population as a whole, where everyone without internet access is excluded right from the outset.

1968 Of Mites and Men

Experiments that are a real struggle fall into two categories: those that earn the researcher the lifelong admiration of his fellow human beings, and those that condemn him for all time to be regarded as a fruitcake. The true heroes of science are to be found in the second category – for example, the vet Robert A. Lopez from Westport, New York.

Lopez once treated a cat twice for ear mites. At the same time, the cat's owner and her daughter complained of itching.

Was it therefore possible for the ear mite Otodectes cynotis to cross over to humans? Scientific literature had nothing to say on the subject, so Lopez decided to run a test of his own. He removed the mites from a cat's ear, verified under the microscope that they were indeed Otodectes cynotis and proceeded to introduce one gram of earwax, with mites mixed in, into his left ear. The effect wasn't long in coming, as Lopez reported: 'Immediately, I heard scratching sounds, then moving sounds, as the mites began to explore my ear canal. Itching sensation then started, and all three sensations merged into a weird cacophony of sound and pain that intensified from that moment, 4 p.m., on and on...'.

Lopez gained an intimate insight into the life of the ear mite: 'The sound in my ear (fortunately I had chosen only one ear), were becoming louder as the mites travelled deeper toward my ear drum. I felt helpless. Is this the way a mite-infested animal feels?' Lopez was also dismayed to note that the ear mites' feeding habits didn't chime in with his pattern of sleep: 'After retiring about 11 p.m., the mite activity increased incrementally so that, by midnight, the mites were very busy, biting, scratching and moving about. By 1 a.m., the sounds were loud. An hour later the pruritus was very intense. After two hours the highest level of itching and scratching was attained.' This pattern repeated itself night after night, and 'made sleep, no matter how demanding, completely out of the question'. Nevertheless, Lopez stuck to his guns.

'By the third week, the ear canal was filling up with debris, and hearing from my left ear was gone. By the fourth week, mite activity was 75 percent reduced and I could feel mites crawling across my face at night.' When his ear became completely blocked with residue, he rinsed it out with warm water, and two weeks later – now free of mites – he was able to hear normally once more.

Yet Lopez wouldn't have been a true researcher if he'd called it a day there and then. If a scientific experiment isn't repeated, then its findings must be regarded as unverified. So, as Lopez later recalled: 'I decided to try again to see whether the first experiment had been flawed or misleading.' Lopez took mites from another cat and put them in his left ear again. The mites initially behaved much as they had in the first test, but after two weeks displayed no activity. This raised a number of questions, Lopez reckoned. Had he perhaps become immune after the first

madsciencebook.com

The Ig Nobel Prizes are awarded every year by the magazine *Annals of Improbable Research* for examples of offbeat research. The associated website, www.improbable.com, contains a rich store of scientific eccentricities.

Source

Lopez, R. A. (1993), 'Of Mites and Man.' *Journal of the American Veterinary Medical Association* 203 (5), pp. 606–607.

experiment? Or were human ears not really a suitable habitat for Otodectes cynotis? And so, 'A third and final trial had to be done.' Once again, the symptoms were less acute. Maybe there really was an immune reaction against the mites, Lopez surmised.

And indeed, after he'd finished the experiment, he did find one such case cited in the scientific literature: a woman who complained of tinnitus brought on by mites in her ear. As Lopez commented in the final sentence of his report on the experiment: 'I wonder whether the person involved enjoyed her experience as much as I did.'

In 1994, Lopez's work earned him the Ig Nobel Prize [in Entomology] for scientific experiments 'that cannot or should not be reproduced'.

1968 Eight Flew over the Cuckoo's Nest

The preparations for this experiment were always the same: David Rosenhan, Professor of Psychology at Stanford University, stopped brushing his teeth for several days. He also didn't wash and gave up shaving. Then he put on dirty clothes, fixed an appointment over the phone with a psychiatric clinic under the false name of David Lurie and had his wife drop him off outside the main entrance.

In reception, he complained of having heard voices, which, insofar as he could make them out, had said the words 'empty', 'hollow' and 'thud' to him. He asked to be admitted to the clinic. The psychiatrist who examined him couldn't have known that Rosenhan had chosen these symptoms very carefully, on the grounds that there was no case in the scientific literature that exactly corresponded to them. But on being admitted, Rosenhan immediately stopped feigning the symptoms. He behaved completely normally, chatting with the patients and staff and biding his time. How long would it be, he wondered, before he was given a clean bill of mental health and discharged? In fact, what he discovered was to spark a major crisis in traditional psychiatry.

Rosenhan was 40 years old when, in 1968, he set about trying to answer the question of whether there were states of 'sanity and insanity' and how one could distinguish the two. 'The question is neither capricious nor itself insane,' he later wrote in his famous article 'On Being Sane in Public Places.' 'However much we may be personally convinced that we can tell the normal from the abnormal, the evidence is simply not compelling.'

By playing the role of a patient, the psychologist David Rosenhan discovered how long a sane person could be kept incarcerated in a psychiatric clinic.

Certainly, the diagnostic handbook of the American Psychiatric Association placed patients in various categories according to their symptoms; these categories were supposed to enable practitioners to tell the difference between mentally ill people and sane ones. Yet Rosenhan had become convinced that a mental illness was less a question of objective symptoms than of the subjective perception of the observer. He believed he could settle the matter by testing whether normal people, who had never suffered the symptoms of a serious mental illness, would appear sane in a psychiatric clinic, and if so, how?

So, between 1968 and 1972, he and seven of those who regularly attended his seminars had themselves committed, under false names and with the same feigned symptoms, to a total of 12 psychiatric clinics. Among the pseudopatients were a psychology student, three psychologists, a paediatrician, a psychiatrist, an artist and a housewife, who were all charged with the task of getting themselves released from the clinic without any external help, by convincing the staff that they were in fact sane. And so they showed themselves to be very cooperative, observing all the clinic's rules and taking all the medication they were prescribed – at least ostensibly: Rosenhan had shown them beforehand how to wedge pills underneath their tongues instead of swallowing them. In total, they were given 2100 pills, including the most diverse concoctions of drugs – all for identical symptoms.

The dangers that his pseudopatients were exposing themselves to only started to dawn on Rosenhan once the experiment was up and running. For instance, some feared that they might be raped or beaten, and Rosenhan realized that he had no possibility of extracting them from the clinics in an emergency. From that moment on, there was always a lawyer on call. Also, since hardly anyone knew about the experiment, Rosenhan left instructions about what to do in the event of his death.

Source

Rosenhan, D. (1973), 'On Being Sane in Insane Places.' *Science* 179, pp. 250–258.

Another fear shared by all the pseudopatients was that they would be instantly exposed. Therefore, at the outset, they kept their research journals in secret. Through an ingenious system, this material was smuggled out of the ward on a daily basis. However, it soon turned out that no special precautions were necessary, as the staff paid no attention.

In fact, not one of the pseudopatients was unmasked. Although they were all discharged in the end, it took three weeks on average for them to be released, and even then they weren't designated as having been cured but rather, in most cases, were diagnosed as 'schizophrenics in remission'. On one occasion, Rosenhan even waited 52 days for his discharge: 'That was a long, long time,' he recalls nowadays, 'but I'd really got used to institutional life.'

Ironically, it was the other patients who saw through the subterfuge. During the first three clinic visits, a third of them voiced their suspicions that the pseudopatients weren't really ill, and some of them really hit the nail on the head: 'You're not crazy. You're a journalist, or a professor [referring to the continual note-taking]. You're checking up the hospital.'

The experiment revealed how deeply ingrained a pigeonholing mentality was in contemporary psychiatry. Once a pseudopatient had been categorized as a schizophrenic during the examination prior to admission, nothing that he did thereafter would rid him of this stigma. Inadvertently, a patient's case history was skewed in such a way that it always ended up confirming the original diagnosis. Classifying someone as mentally ill also meant that the clinic's staff overlooked quite normal behaviour or misinterpreted it. For instance, the notes on one pseudopatient who was busy at the time writing up his research journal ran: 'Patient engages in writing behavior.'

Rosenhan and his fellow pseudopatients also conducted a few little experiments on the clinic staff. From time to time, they would ask the nurses and doctors for permission to go outside and then observed what happened next. The most common reaction was for the person concerned either to answer in passing with their gaze averted or not to respond at all. These encounters often ran along exactly the same lines:

Pseudopatient: 'Pardon me, Dr X, could you tell me when I will be eligible for grounds privileges?'

Doctor: 'Good morning, Dave. How are you today?' (The

doctor then continued on his rounds without waiting for an answer.)

Around the same time, the theme of patients in psychiatric clinics being incapacitated was also approached from a quite different direction: in 1962 the hippie author Ken Kesey published the novel *One Flew Over the Cuckoo's Nest*, which was turned into a hit film in 1975 with Jack Nicholson in the leading role. In it, Nicholson plays the petty criminal Randle Patrick, who has himself committed to a psychiatric clinic to avoid being sent to jail.

This book could very well have been the inspiration behind the experiment, for the reader is constantly confronted with the question of who exactly the mad people are here, the inmates or the staff. However, according to his own account, Rosenhan hadn't heard of *One Flew Over the Cuckoo's Nest* when he embarked on his experiment in 1968.

madsciencebook.com **Watch David Rosenhan talk about his attempts to get amitted into a psychiatric hospital – and to get out again.**

The publication of the experiment in 1973 unleashed a storm of protest. Several of Rosenhan's colleagues criticized the study for its supposed methodological shortcomings, while others regarded 'schizophrenia in remission' as tantamount to 'sane'.

Despite the criticism levelled at Rosenhan's study, it still created considerable ripples. Rosenhan had never denied that certain behaviours deviate from the norm, or that people suffer from hallucinations, anxiety or depression. But he did think that the rigid classification of the diagnoses of these conditions was at best highly equivocal and at worst actively harmful. Though nobody actually did away with classification in psychiatric diagnosis after his study was published, lists were drawn up enumerating types of behaviour that had to be present in order to categorize a condition as a particular illness. Yet the destigmatizing of certain diagnoses such as 'schizophrenic' or 'mentally ill' still hasn't happened. People appear to be abnormally strongly influenced by predetermined classifications of disorders. If a person is seen as mentally ill, then everything she does is interpreted in this context.

In another, extremely elegantly designed experiment, Rosenhan demonstrated that this attitude of preconceived expectation also operated in the converse situation. The catalyst for this 'non-existent impostor' experiment came when the staff at a clinic, who had heard of his earlier experiment, claimed that such misdiagnoses couldn't have occurred at their institution. Rosenhan suggested the following test to them: over the next three months, he would send them one or more pseudopatients, giving them the chance to test their expertise in sniffing them out.

Over the three months, the clinic in question admitted 193 patients. Nineteen of them were identified by a psychiatrist and another member of staff as being potential pseudopatients. The only problem was that Rosenhan hadn't sent them any pseudopatients during this period.

1969 The Vandal in Us All

On his regular drive to work, the psychologist Philip Zimbardo (see p. 214) had ample opportunity to study the subject of vandalism in New York. On a single day, for instance, on the 19-mile (30-km) route between his place of work, New York University in the Bronx, and his home in Brooklyn, he counted 218 cars wrecked by vandals.

How did such wanton acts of destruction come about? Zimbardo devised a test to find out: together with a colleague, he purchased a ten-year-old

Zimbardo's magnet for vandalism and looting in the Bronx. After 26 hours, hardly anything of the experimental car remained.

Oldsmobile, which he parked opposite the university campus. From his observations, he knew that some kind of trigger was needed to set the wrecking process in motion. So, he took off the car's number plates and propped open the bonnet before retiring to his observation point. Twenty-six hours later, a procession of looters had made off with the battery, the radiator, the air filter, the antenna, the windscreen wipers, the right-hand side chrome trim, all the hub caps, the jump lead, the spare petrol can, a can of silicone wax and the left-side rear tyre – the other tyres being too bald to make it worth anyone's while stealing them. The first looters – a couple with their eight-year-old son – got to work just ten minutes after Zimbardo had left the vehicle. The mother kept lookout while her son passed his father the tools he needed to take out the battery. The whole operation took them seven minutes.

madsciencebook.com **The full text of Kelling and Wilson's famous article 'Broken Windows'.**

The destruction of the Oldsmobile followed a pattern that was already familiar to Zimbardo from his studies: the first items to be lifted were all those things that could be reused or sold. When there's nothing usable left, kids and youths take possession of the car, smashing its headlights and windows. Then the bodywork is pelted with bricks and bashed with hammers and steel piping, until the vehicle finally becomes a public rubbish tip.

In less than three days, then, the car had been reduced to a useless heap of metal by '23 incidents of destructive contact'. It was often the case that passers-by would stand and watch the vandals at work, and, contrary to Zimbardo's expectations, the destruction took place in broad daylight.

At the same time, Zimbardo had also left a plateless automobile with an open bonnet by the side of the road in the university town of Palo Alto in California. There, however, no harm befell the car. When it started to rain, a passer-by even slammed the bonnet shut. Zimbardo tried again, this time parking the car on the university campus itself. Again, nothing happened.

Yet he was convinced that the citizens of Palo Alto also had it in them to be vandals: 'It was obvious that the releaser cues which were sufficient in New York were not adequate here.' To help things along a bit, Zimbardo and two of his students picked up sledgehammers themselves and set an example. And, for sure, it wasn't long before other students were joining in. They jumped on the car roof, ripped the doors off their hinges, broke all the windows and finished up by tipping the car onto

Source

Zimbardo, P. G. (1969), 'The Human Choice: Individuation, Reason, and Order versus Deindividuation, Impulse, and Chaos.' In: *Nebraska Symposium on Motivation* 17 (Arnold, W. J., ed.), University of Nebraska Press. pp. 237–307.

its roof. In the middle of the night, three teenagers then appeared and attacked the wreck with iron bars.

Clearly, in Palo Alto, either the cover of darkness or the anonymity of a group was needed to awaken dormant vandalistic tendencies. The threshold appeared to be far lower in the Bronx in New York. Zimbardo assumed that the anonymity of the big city and the signs of general decay in the run-down neighbourhood where the car was parked increased people's tendency to destructive behaviour.

The criminologist George L. Kelling and the political scientist James Q. Wilson used these findings to construct one of the most far-reaching theories in the history of criminology. In the March 1982 edition of the scholarly American publication *Atlantic Monthly*, under the title 'Broken Windows', they published an article in which they put forward a new strategy for combatting criminality. In it, they claimed that the best way of doing this was to focus on the acts of disorder that preceded it: 'One unrepaired broken window is a signal that no one cares, and so breaking more windows costs nothing.'

From their own experiments and surveys, Kelling and Wilson knew that people were troubled by minor antisocial acts, such as graffiti, litter on the street and vandalism. These made them feel as though things had got out of hand, and that nobody was taking responsibility for anything any more. And it was this feeling that prepared the ground for serious criminal acts: on the one hand, citizens and even to some extent the police withdraw from public spaces, leaving them free to become lawless zones, while on the other, wrongdoers' inhibitions at committing further, more serious criminal acts were steadily eroded.

The idea that this process could be reversed by combatting its outward signs was greeted with scepticism. People thought that criminality could really only be tackled effectively by getting to its roots. And that, depending on one's political outlook, was to be found either in social injustice or in society's moral decline.

In the 1990s, New York police chief Bill Bratton implemented the 'broken windows' strategy in the Big Apple. Graffiti-sprayed subway trains were immediately withdrawn from service and cleaned, drunks and beggars were cleared off the streets and litter was tidied up. Since Bratton took office in 1994, the murder rate in New York has been almost halved. But whether this success really can be attributed to his zero-tolerance policy remains, as might be expected, the subject of much debate.

1969 The Ape in the Mirror

It's one of the oldest questions posed by science: do animas possess a sense of self-awareness? For a long time, no one could come up with a method of answering this, and many researchers became convinced that the problem was basically insoluble. The general wisdom was that consciousness simply wasn't a sensible subject for investigation. A psychologist at the time summed up the situation: 'Unfortunately, there is no way to interview animals to discover the exact point on the evolutionary scale at which consciousness emerges. Neither is there any way to determine when "self" becomes an element within the subjective map.'

madsciencebook.com Watch the mirror experiment recently conducted with an elephant.

The simplest instrument for self-perception is the mirror. Even Darwin experimented with it. 'Many years ago, in the Zoological Gardens, I placed a looking-glass on the floor before two young orangs, who, as far as it was known, had never before seen one,' he wrote in his 1872 work *The Expression of the Emotions in Man and Animals*. The apes' response was extremely animated. They wanted to kiss their reflection, pulled faces and looked behind the mirror. Later, they refused to look at the mirrors any more.

Did the apes take their reflection to be another ape, or did they recognize it as themselves? Darwin couldn't answer this question. Sure, he could observe their behaviour but anything he might have concluded from it was based solely on his subjective interpretations and so did not constitute a scientific proof.

All the researchers who came after Darwin also foundered on this problem – until Gordon G. Gallup suddenly had a devastatingly simple brainwave while he was shaving. The year was 1964, and Gallup was studying for his doctorate at Washington State University. Five more years elapsed – during which time he became a professor – before he was able to put his idea into practice at Tulane University in New Orleans.

Like Darwin, his method was to show apes their reflection in a mirror. In separate rooms, four young chimpanzees were placed in cages, in front of which Gallup placed large mirrors. He sat and observed their behaviour for ten days. Would the chimps realize that the mirror was a source of information about themselves? For the first day or so, they reacted to their reflection as though it were a stranger: they threatened it and shrieked, and ran away and attacked by turns. This was to be expected; after all, they had never seen an ape in a mirror before. It is also the case that human children under the age of two don't recognize their reflections as themselves. However, on the third day, the chimps' behaviour changed: they stood in front of the mirrors cleaning out bits of food that had got stuck between their teeth or picking lice off places on their coats that were otherwise invisible. Gallup was sure that the apes were now reading their reflection correctly, yet he was still no further on than Darwin. His conviction amounted to no more than a personal opinion. Anyone claiming that a chimp had a sense of self-awareness would have to be able to come up with hard-and-fast proof. But, thanks to a trick, Gallup was able to supply precisely this.

On the tenth day, he tranquillized the chimpanzees and painted one spot of red dye on one eyebrow and another on the ear opposite. They could neither smell nor feel the dye; Gallup had tried it out on himself a few days earlier.

Once the chimps had come round from the anaesthetic, Gallup counted how often they touched the red spots, first without the mirror and then with. The result was unequivocal: with the aid of the mirror, they touched them 25 times more frequently (without the mirror, it was evident that they only touched them by chance). In order to be quite certain, Gallup also applied paint in the same way to chimps who had never seen a mirror before. These animals exhibited no response whatsoever.

This meant that, at some point during the ten days, the chimps must have realized the significance of their own reflection. As Gallup later said, 'Man may not be evolution's only experiment in self-awareness.'

Thereafter, Gallup's dye-mark test was conducted on every possible kind of animal, and is now often the first scientific experiment that animal behaviourists perform with their own children (mostly by surreptitiously sticking a Post-It note on their forehead). But the only creatures, apart from humans aged around two upwards, to pass the test are chimpanzees, orang-utans, a single gorilla who had grown up among humans and, in 2005, an elephant called 'Happy' from the Bronx Zoo. Maxine and Patty, the other two elephants from the same enclosure, didn't make the grade. In addition, dolphins and magpies are supposed to be able to recognize themselves in mirrors. With dolphins and magpies, though, the results are contentious; lacking hands with which to touch the paint dot, Gallup's method is of only limited use where they are concerned.

Source

Gallup, G. (1970), 'Chimpanzees: Self-Recognition.' *Science* 167 (3914), pp. 86–87.

Nobody will ever be able to say with certainty whether it was pure chance that Gallup had his brainwave while shaving, in catching sight of the white blotches of shaving foam on his face in the mirror. Gallup himself isn't convinced that this situation played a key role, but on the other hand such thought processes rarely occur consciously.

Despite the success of the test, it's hard to say exactly what it measures. What does it signify if an animal has the capacity to recognize itself in a mirror? In doing so, is it instantly certain that the reflection is itself? Does this mean that it can put itself in another's position? Or recognize others' intentions? Or lie? – all things that are part and parcel of self-awareness.

Gallup's dye-mark test is a good example of the fact that science can often provide interesting answers without precisely knowing what the question is.

1969 Colour Tests in the Primeval Forest

When Eleanor Rosch crossed over the border from Papua New Guinea to Irian Jaya (now West Papua) in the summer of 1969, the frontier guards didn't have a clue what to make of the hundreds of playing-card-sized colour cards that they found in her luggage. Rosch didn't even try to begin to explain to them that she intended to use the cards to disprove a contentious theory in linguistics. So, after giving a few vague responses and showing their official travel documents, she and her then-husband, the anthropologist Karl Heider, were allowed through the checkpoint.

At the time, Rosch was studying for her doctorate at Harvard University; it was there she met Heider, who told her about a peculiarity of the Dani people. He had already gone on several fieldtrips to visit these hunter-gatherers and had discovered that they had only two words for colours: 'mili' for dark colours, and 'mola' for light ones. Rosch realized straight away that these might provide the key to solving an age-old problem in language research: namely, how does language influence thinking?

In the 1930s, the linguist Edward Sapir inclined towards the view that language determined thought. Language didn't adapt itself to reality, rather it was the case that reality could only be perceived through the lens of one's own native language. In other words, every language embodied a different worldview. This implied, then, that reality wasn't something objective out there in the physical world but that it resided in every individual's head and was furnished with all the trappings of one's mother tongue.

In reference to Einstein's special theory of relativity, Sapir's pupil Benjamin Lee Whorf called this principle 'linguistic relativity'. Later it also became known as the Sapir–Whorf hypothesis. Anyone who followed this idea right through to its logical endpoint might conclude that two peoples speaking different languages could never really understand one another.

Whorf thought that he'd chanced upon proof of his hypothesis in the language of Native American peoples: for example, the Hopi had only a single word for everything other than birds that can fly, while the Eskimos had seven different words for snow. Whorf also thought that his idea was confirmed by grammar: since there are no temporal forms of verbs in the Hopi language, he assumed that they had a different conception of time. However, the examples he cited were prone to circular logic: Whorf deduced a different worldview from the peculiarities of language. But he could equally have expressed it the other way round: because the world of the Native Americans was different, then they spoke about it differently.

Initially this dilemma seemed insoluble, unless it were somehow possible to observe thought and perception independently of language and categorize them against an objective yardstick. Yet there's no way of measuring a worldview objectively or of communicating it independently of language. In other words, there is no hard-and-fast physical scale on which to place the supposedly different conception of time among the Hopi.

The solution to the problem was colours: colours can be categorized according to their wavelengths, quite independently of subjective perception. And how a test subject subdivides a palette into its individual colours can be determined independently of language. Now all that was needed were people who had a large number of different words for colours in their native language in order to test whether these linguistic differences also meant that they saw the world in different colours.

The first experiments in the 1950s yielded no clear results, but at the end of the 1960s Brent Berlin and Paul Kay from the University of California at Berkeley compared over 100 languages and found that colour words developed in various different languages according to a set pattern. If a language only had two

The psychologist Eleanor Rosch conducting an experiment in Papua New Guinea that focused on the recognition of facial emotions. She ran this experiment concurrently with her colour tests.

words for colours, then these were invariably 'black' (for all dark colours) and 'white' (for all light ones); if there were three, then they were 'black', 'white' and 'red'; if four, 'black', 'white' and 'red' and either 'yellow' or 'green'. One by one, each of the 11 basic colours was added to the list. This principle indicated that colour perception obeys certain universal rules.

A test also pointed in the same direction: Berlin and Kay presented speakers of 20 different languages with a palette of colour cards and asked them to demarcate from one another various colours for which different names existed in their language. They were also required to pick a typical shade to represent each different named colour.

Although the test subjects drew the boundary between individual colours at different points, their choice of typical shades was nevertheless very similar. Therefore, it seemed that there were a number of focal colours that were commonly perceived irrespective of language and culture.

However, this still didn't comprehensively scotch the notion that language governs perception: the test subjects were immigrants who had already been exposed to the influence of English for some time. Only tests on subjects who had had hardly any contact with other language communities would be truly meaningful.

The Dani were just such a people. In the summer of 1969, Rosch carried out her first experiment with 40 men at Yibika, a Dutch missionary outpost. She showed each of them a colour card from a stack for five seconds, and then waited 30 seconds before asking them to pick the card they had seen from a set of 40 cards, laid out according to brightness and shade. She repeated this process for all the cards in the stack. At the same time, she counted how often a subject went wrong and, instead of lighting on the correct card, pointed to an adjacent one. Her reasoning was simple: if Whorf was right and language influences perception, then the Dani would be more likely to confuse colours for which they only had a single word than those for which there were several words in their language.

Source

Rosch Heider, E. (1972), 'Universals in Color Naming and Memory.' *Journal of Experimental Psychology* 93 (1), pp. 10–20.

However, a comparison with a group of American students who had also taken the same test showed no such effect. For instance, the Dani had no more trouble distinguishing colours on the borderline between blue and green than Americans, despite the fact that they had only one word –'mola' – for such colours. This seemed to refute the Sapir–Whorf hypothesis.

Yet this was really only the start of a long-running debate, for lurking behind the question of whether a person sees the world through different eyes just because they happen to speak a different language was another one, namely: to what extent is the human mind shaped by our environment? The seemingly harmless colour test could be used to distinguish environmental from genetic influences.

Passions run high in this debate, as the whole nature–nurture question is a hot potato, especially in the realm of gender politics. For instance, does the use of gender-specific nouns in many languages to denote particular professions skew from the outset boys' and girls' choice of career?

It was predictable that there would never be any agreement on this matter. In 1999, the results of a study were published that refuted Rosch's findings: Debi Roberson from the University of London repeated Rosch's experiment with another people on New Guinea, the Berimo. She concluded that the fact that they had only five words for colour did indeed influence their colour perception. Roberson takes the view that there are methodological flaws in Eleanor Rosch's experiment. For her part, Rosch believes that Roberson's choice of colour cards is inappropriate.

In any event, the Eskimos' words for snow have become an urban myth. The linguist Franz Boas, who first cited this example, found four different words for snow in the Eskimo language in 1911. Whorf increased this number to seven, and journalists then inflated this figure to such an extent that a weather report for Cleveland, Ohio, ended up talking about 100 words for snow. Experts now think that a more realistic figure is around a dozen.

1970 How Embarrassing!

When Howard Garland spotted the open toilet window, he knew that his experimental method had been a success. Garland, a student at Cornell University in New York, was working on the psychology of embarrassing situations, tackling such questions as why things are embarrassing in the first place and how people try to save face. To answer these, he had to create a situation in the laboratory that the test subject found embarrassing. In earlier experiments, Professor Bert R. Brown had got his students to suck on a dummy and describe their feelings in public. 'In this case,' Garland recalled, 'I was the public and even I found the whole thing embarrassing.' The dummy method had been devised in the early 1960s and had a strong sexual undertone. Although it was extremely effective, Garland wondered whether there wasn't a gentler way of embarrassing a test subject. And so he hit on the idea of getting people to sing a song in public.

Quite how effective this approach was was brought home to him when one of the participants in the experiment asked him where the loo was just before he was due to sing. Garland gave him directions and waited, but the student didn't reappear. When

Garland went to check on him, the toilet – which was on the first floor – was empty and the window was open. The student who'd done a bunk hadn't hung around to discover that the experiment's guidelines would have enabled him to withdraw without being forced to sing anyway.

To conceal the true purpose of the experiment, Garland disguised it as a test for a new music computer that was programmed to judge people's singing voices. The way that this worked, he explained to the test subjects, was that their singing would be judged twice over, once by the computer and once by a live audience. A comparison of the two scores would demonstrate how closely the computer's assessment mirrored that of real people.

Source

Brown, B.R. und Garland, H. (1971) The Effects of Incompetency, Audience Acquaintanceship, and Anticipated Evaluative Feedback on Face-Saving Behavior. Journal of Experimental Social Psychology 7, pp. 490–502.

After this introduction, the participants were played a tape of the song 'Love is a Many-Splendored Thing', a cheesy ballad from the 1950s, which Garland had chosen for its high 'embarrassment potential'. The song required an extensive vocal range and its lyrics were pure kitsch. The test subjects were allowed a short time to practise before performing it in front of the supposed computer, which then gave its verdict: 'good' or 'poor'.

Garland then showed the subjects into a small room with a one-way mirror, behind which he told them an audience was sitting (in fact, the only person there was Garland himself, with a stop-watch). He had told them beforehand that they would get a cent for every five seconds they sang. The point at which they stopped singing was a clear indicator of how embarrassing they found the whole situation.

The first results were as expected: if a person's voice had been judged 'poor' by the computer, they sang on average for 82 seconds, whereas a 'good' voice managed 132 seconds (the computer verdicts were completely random; after all, there was no computer in reality). The test subjects also sang for a noticeably shorter time if they thought that friends of theirs were behind the mirror than if they imagined strangers there. When Garland tested men and women separately, he got a surprising result: women sang on average for just 16 seconds in front of a female audience, whereas they lasted four times as long in front of a male one. The reason for this, he believed, was that women generally reckoned men to be worse singers, who wouldn't notice the odd bum note when listening to someone else.

1970 The Bad Samaritans

The psychologists John M. Darley and C. Daniel Batson got the idea for this experiment from the Bible (in the Gospel According to St Luke):

'A man was going down from Jerusalem to Jericho, when he fell into the hands of robbers. They stripped him of his clothes, beat him and went away, leaving him half dead. A priest happened to be going down the same road, and when he saw the man, he passed by on the other side. So too, a Levite, when he came to the place and saw him, passed by on the other side. But a Samaritan … went to him and bandaged his wounds, pouring on oil and wine. Then he put the man on his own donkey, took him to an inn in Jericho and took care of him.'

When Batson looked closely at the parable of the Good Samaritan, he found three prognoses regarding people's willingness to help a fellow human being:

First: anyone in a hurry is less likely to help. The priest and the Levite, Batson thought, were religious functionaries 'hurrying along with little black books full of meetings and appointments, glancing furtively at their sundials.' The Samaritan, on the other hand, wasn't a high roller and had time to help.

The 'victim' in the doorway. Was he helped by theology students, who had just been preparing a lecture on the subject of the Good Samaritan?

Second: anyone who's busy mulling over ethical or religious questions when their help is called for isn't any more likely to help than someone who's thinking about something else. The priest and the Levite must often have had call to dwell on religious matters, and were in all likelihood doing so at the very moment when they came across the victim. By contrast, the Samaritan was probably thinking about more worldly things.

Third: anyone who's devout because he hopes it will bring him personal gain lends his help less readily than those who regard religion as a constant quest for meaning in their daily lives, without any ulterior motives. The Levite and the priest belonged in the first category, the Samaritan in the second.

Batson wanted to test these three hypotheses. But to do so he needed religious test subjects. 'This means that the standard college sophomore subject pool may be less than totally appropriate,' he noted ironically. Batson had another group in mind: students from the Theological Seminary at Princeton University.

He had also already chosen his 'road from Jerusalem to Jericho': an asphalt path between the Psychology Department and the back door of an adjacent building on the Princeton campus. Although this alleyway wasn't frequented by robbers, it was dark, shabby and off the beaten track. Batson asked the few people who regularly used it to use another route for the duration of the experiment.

At 10 o'clock on the morning of 14 December, Batson sent the first theology student on his way. The student couldn't have had any inkling that he was 'going down from Jerusalem to Jericho', since Batson had disguised the experiment. The subjects came to the Psychology Department in the belief that they were taking part in a survey on religious education and their vocation. There they were asked to prepare a lecture of three to five minutes' duration, which they were then supposed to record on tape. Because it was too crowded in the Psychology Department, they were sent across on the chosen route to Sociology, where an assistant was waiting with the tape recorder.

It was on this route that they came across the 'victim', a man with tousled hair and hands stuffed deep into the pockets of his anorak, who was sitting slumped and with his eyes closed in the doorway of the adjacent building. The man playing the victim

Source

Darley, J.M. und Batson, D. (1973), 'From Jerusalem to Jericho: A Study of Situational and Dispositional Variables in Helping Behavior.' *Journal of Personality and Social Psychology* 27, pp. 100–108.

obeyed the instructions he had been given to the letter, coughing twice and moaning as each subject drew near. If they asked whether he was all right, he replied, 'Oh, thank you [cough] .... No, it's all right. [Pause] I've got this respiratory condition [cough] ...The doctor's given me these pills to take, and I just took one ....If I just sit and rest for a few minutes I'll be O K ...Thanks very much for stopping though [smiles weakly].' If the subject insisted on taking him into the building, he complied.

The question was under what conditions the victim was offered help. Some of the subjects were encouraged to make haste while they were still in the Psychology Department: 'Oh, you're late. They were expecting you a few minutes ago. We'd better get moving. The assistant should be waiting for you so you'd better hurry. It shouldn't take but just a minute.' Others, though, were told they were running early and had plenty of time to get over to the adjacent building. Batson also varied the theme of the lecture that the subjects were meant to deliver: half of the subjects were to give a three-minute talk on favourite careers for graduates of the Theological Seminary, while the others had to speak about the parable of the Good Samaritan. Batson probed the third potential factor influencing their behaviour – what tenets of faith the subjects shared – by means of a questionnaire.

Over the course of three days, he sent 47 students over to the adjacent building. And not without incident. One type among the subjects was a kind of 'super-helper': people who simply wouldn't let up until they'd sat the victim down over a cup of coffee and bent his ear about the love of Jesus. This threw a spanner into the planning of the experiment, which called for a new subject to be dispatched every half an hour.

The result was surprising: the only circumstance that affected a person's willingness to help was how rushed they were. Test subjects with time on their hands offered to help six times more frequently than those who were in a hurry. People's personal creeds didn't yield a clear result, though dogmatic theology students certainly were prominent among the super-helpers. But the most astonishing result related to another matter entirely: whether people helped or not had nothing whatever to do with them thinking at the time – or not, as the case may be – about the parable of the Good Samaritan. In fact, there were several test subjects who didn't bat an eyelid about stepping over the victim to go and give their talk about the inhumane conduct of the priest and the Levite.

After the students were told about the real background to the experiment, many were ashamed and subsequently used their own behaviour as instructional material in their sermons.

Batson intended to conduct more experiments in the same location with larger numbers of participants. But just as he was preparing the new set of tests, the asphalt alley with its puddles and rubbish bins fell victim to a drive by Princeton University to tart up its grounds. So now, in its place, there stands a pleasant, secluded path, complete with a bench and a tree, leading to the adjoining building. 'Everyone agrees that it is a much prettier place, and of course it is. But it's hard not to regret the loss of a beautiful sinister alley,' Batson later wrote.

1970 The Dollar Auction

In 1970, when Allan I. Teger was discussing the psychology of international relations with his students at the University of Pennsylvania, there was no shortage of illustrative material. The USA was fighting the Vietnam War at that time, and every day the papers were full of the kind of themes that Teger's course focused on: decision making, retribution and group dynamics.

One argument repeatedly trotted out by the US administration made a particular impression on Teger, since he believed that it represented a significant cause behind the escalation of the conflict: even though the costs of the war would never outweigh its benefit, the USA would keep on fighting so that 'our dead should not have died in vain.' Put another way, the country had already pumped too much money into the war to give up on it now.

Teger pondered how to really bring the workings of this logic home to his students. Although there were mathematical games that simulated conflicts, such as the prisoner's dilemma (see pp. 129–132), the players remained largely calm in these. They don't get furious or frustrated. Teger had something more realistic in mind, and so hit on the idea of an auction with its own peculiar set of rules. Shortly afterwards, he read that the economist Martin Schubik had already thought up a similar game, and he borrowed from Schubik the particular lot up for sale, namely a dollar bill.

Just like in a normal auction, the dollar bill is put up to be sold to the highest bidder. But in Teger's version, there was a devilish additional rule, whose effect most participants only realize when it's already too late: the person who bid the second highest amount also has to pay up, without receiving anything in return. Everyone else who made lower bids pays nothing.

Teger auctioned his first dollar bill in a lecture. To start with, everybody joined in the bidding. No-one wanted to miss out on the chance of getting a dollar for less than a dollar. But when the bidding had got up to around 70 cents, most people began to realize how treacherous the supplementary rule was and dropped out of the running. The only people left were the two students with the highest bids, who were already caught in an impossible situation. The first bidder had bid 80 cents, the second 90. If the first bidder dropped out now, he'd have to pay 80 cents and have nothing to show for it. He could only prevent this by bidding a dollar. Although this meant he won nothing – he was buying a dollar for a dollar, when all was said and done – at least he wasn't losing anything. Now the second bidder was on the horns of a dilemma: if he dropped out, he'd lose 90 cents. In these circumstances, it was better to go up to one dollar and ten cents. When this bid was made, a murmur ran through the lecture theatre. How could anyone be buying a dollar for $1.10? He'd be losing 10 cents! But if he hadn't put in a bid, his losses would have amounted to 90 cents. And so the thing escalated, as the two were forced to outbid one another.

Teger did this experiment about 40 times, and for every dollar that he auctioned, he never failed to attract bids of more than a dollar – sometimes as much as $20. He never collected the money, though. The only essential thing about the experiment was that, during the auction, the players should believe that they'd have to pay up.

When Teger canvassed the test subjects' thoughts after the auction, many of them excused themselves for their irrational behaviour. Business students found it especially embarrassing that they'd lost money. One even tried to explain away his behaviour by claiming that he'd been drunk.

Yet their behaviour had been completely normal: it was the rules of the game that lured them inevitably to their ruin. The same thing applied to the Vietnam War: 'It was not just a bunch of idiots who say we are gonna kill people. These are people trying to get

Source

Shubik, M. (1971), 'The Dollar Auction Game: A Paradox in Non-cooperative Behavior and Escalation.' *Journal of Conflict Resolution* 15, pp. 109–111.

Teger, A.I. (1980), *Too Much Invested to Quit*, Pergamon Press.

themselves out of the mess.' But in trying to extricate themselves they only succeeded in digging themselves deeper.

In his interviews with the bidders, Teger also ascertained that there'd been a fateful shift in their motivation. At the start, they were just drawn in by the prospect of a quick buck. But as the bidding approached a dollar, they all found themselves facing the same dilemma: should they stop and lose stake or carry on bidding? However, in many cases the motivation to continue had nothing to do with money any more. Rather, it was now about winning at all costs. And about making the other guy suffer. Most bidders were convinced that their rival bidder was responsible for getting them into this hopeless situation. When asked about what motivated their adversary, some people even responded that he must be crazy. It never dawned on the participants that the game was symmetrical and that the other bidder must be thinking exactly the same thing about them.

The dollar auction is a parable for the escalation of conflicts. The book that Teger wrote about his research came to be deployed in workshops aimed at solving the conflict in Northern Ireland, as well as in disputes between firms. When he himself was faced with the question of whether he hadn't already invested too much to quit, he answered with a decisive 'No': in 1981, he gave up academic research and became a photographer.

1970 Dr Fox Spouts Nonsense

The lecture that Myron L. Fox delivered to the assembled experts had an impressive enough title: 'Mathematical Game Theory as Applied to Physician Education'. Those responsible for running the University of California School of Medicine's further-education programme had taken themselves off to Lake Tahoe in northern California for their annual conference. There, Fox – who was billed as an 'authority on the application of mathematics to human behaviour' – presented the first paper. His polished performance so impressed the audience that nobody noticed that the man standing at the lectern wasn't just Myron L. Fox from the Albert Einstein School of Medicine: he was also the Gotham City radio manager Leo Gore from *Batman*, the attorney Amos Feders from *Falcon Crest* and Dr Benson from

Columbo, the vet who looked after the inspector's dog. Myron L. Fox's real name was Michael Fox, and he was an actor (though no relation of Michael J. Fox of *Back to the Future* fame). And he didn't know the first thing about game theory.

All that Fox had done was to take a scholarly article on game theory and work up a lecture from it that was quite intentionally full of imprecise waffle, invented words and contradictory assertions. Fox delivered this lecture in a very humorous tone, all the while making specious references to other supposed works. The people behind this spoof were John E. Ware, Donald H. Naftulin and Frank A. Donnelly, who wanted to use this demonstration to spark a discussion on the content of the further-education programme. The experiment was designed to find out whether a brilliant delivery technique could so completely bamboozle a group of experts that they overlooked the fact that the content was nonsense. John Ware put in hours of practice with the actor, to the point where the text had been stripped of all its substance. As Ware reported, 'the problem was to keep him [Michael Fox] from making sense'.

Fox was convinced he'd be rumbled. But the audience hung on his every word and, when the hour-long lecture was over, bombarded him with questions, which he displayed such virtuosity in not answering that nobody noticed. And on the feedback form that was handed round, all ten people who attended the lecture said that it had given them food for thought, while nine of them also reckoned that Fox had presented the material in a clear manner, put it across in an interesting way and incorporated plenty of good illustrative examples into his talk.

The brilliant lecturing of actor Michael Fox completely duped the experts.

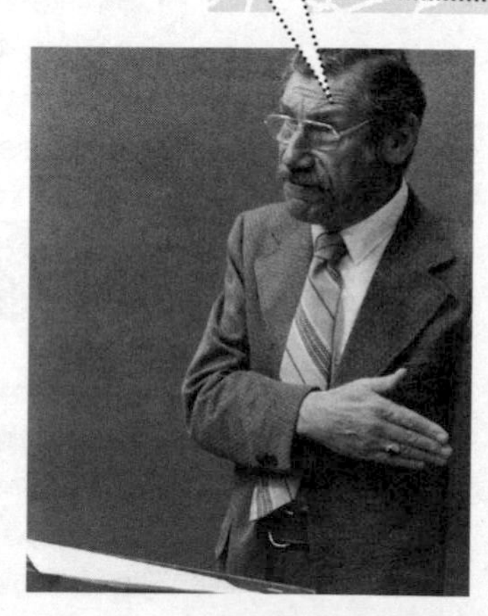

Ware and his colleagues showed two other groups of people a video of the lecture – with much the same result. One person even thought they remembered having read some papers already by Myron L. Fox. In these instances as well, the audience wasn't made up of students but of experienced educators, who had allowed themselves to be dazzled by the actor's slick presentation.

The researchers conducted further experiments on larger audiences. The phenomenon in which the style of a lecture can blind the listeners to its poor content soon became known as the 'Dr Fox effect'.

These results raised doubts in Ware's mind about the usefulness of teaching evaluation. When students were asked to fill out questionnaires assessing a class, these might actually be indicating little more than how much they liked the lecture along with 'their illusions of having learned'. As the authors wrote in their paper on the experiment, 'there is much more to teaching than making students happy'.

Source

Naftulin, D.H., et al. (1973), 'The Doctor Fox Lecture: a Paradigm of Educational Seduction'. *Journal of Medical Education* 48 (7), pp. 630–635.

Nevertheless, there was one surprise that qualified this conclusion: when Fox's true identity was revealed to the audience, some of them asked where they could read up more about the subject. In other words, although the lecture had been unmasked as gibberish and a fraud, the panache with which it was delivered had nevertheless clearly stimulated interest in the topic. This led Ware to suggest an innovative method of increasing students' motivation: instead of giving lectures themselves, professors could train actors to deliver lectures for them.

A journalist later wrote in the *Los Angeles Times*: 'There are implications in this study, though, that even its instigators have not perceived. If an actor makes a better teacher, why not a better congressman, or even a better President?' Ten years after Ware's hoax, Ronald Reagan was elected to the White House.

1971 The Professor's Prison

In his hotel room on the evening of 28 April 2004, Philip Zimbardo switched on his television. At the time, Zimbardo was president of the American Psychological Association and was staying in Washington for a conference. Flicking around the channels, he suddenly found himself transfixed in horror by the image on the screen of naked prisoners piled up into a pyramid. Behind them stood two laughing American soldiers, a woman and a man. The next image he saw was of another female soldier who was holding a prisoner cowering on the floor on a leash. Then came the image that was later to become an icon: a prisoner with a sack over his head and electrical cables attached to his hands, teetering on a small wooden crate. The soldiers had told him that if he fell off the crate, he'd receive an electric shock that would kill him.

In the spring of 1971 psychologist Philip Zimbardo placed this ad in the *Palo Alto Times*.

ity required. No selling Guar hrly. Wage. Write Redwood City Tribune AD No. 499, include phone no.

Male college students needed for psychological study of prison life. $15 per day for 1-2 weeks beginning Aug. 14. For further information & applications, come to Room 248, Jordan Hall, Stanford U.

TELEPHONE CONTACTS Sell security systems. New air cond. office, parking. Exc. location & understanding boss

These images came from the prison of Abu Ghraib in Baghdad, which the US military used for torturing prisoners after their 2003 invasion of Iraq. How could this have been allowed to happen? The administration in Washington was quick to claim that the torturers were just a few rotten apples in the barrel and that regular US troops simply weren't capable of such misconduct. Philip Zimbardo knew straight away that this wasn't true. And he of all people should have known. Thirty years earlier, he had been responsible for setting up a prison where torture took place.

In the spring of 1971 Zimbardo, then aged 38, placed an advertisement in the *Palo Alto Times*: 'Male college students needed for psychological study of prison life. $15 per day for 1–2 weeks beginning Aug. 14. For further information & applications, come to Room 248, Jordan Hall, Stanford U.'

The idea for the experiment came from the course that Zimbardo taught at the university. Some students had chosen the theme 'the psychology of imprisonment' and had played at prisons for a whole weekend. Zimbardo was surprised by the deep impression that this brief experience had made on the students, and decided to investigate the matter further.

Out of the more than 70 applicants who reported to Room 248, Zimbardo chose students who had emerged from a series of personality tests as particularly honest, reliable and stable. He tossed a coin to divide them into groups of guards and inmates. Eleven students were then informed by telephone that they would play the prisoners and were told to remain available at their homes on Sunday 15 August. Ten others, who were to play the guards, were introduced to the 'prison superintendent Philip G. Zimbardo and his deputy David Jaffe – a research assistant – on the day before the start of the experiment. They were shown the prison, which had been set up in the cellar of the Psychology building. Three small laboratory rooms, whose doors had been replaced by bars, were to serve as cells. There were also surveillance rooms for the guards and a 30-ft (9-m)-long corridor, watched over by a video camera, which was used as an area where inspections could take place. Via an intercom system in the cells, the warders could both issue orders and secretly listen in on the inmates.

One of the prisoners in his regulation white smock. The test subjects were assigned at random to be either a prisoner or a guard.

The guards all went together to an army-surplus store to pick out their uniforms – khaki shirts and trousers – and were each issued with a whistle, mirror sunglasses and a rubber truncheon. They worked in eight-hour shifts and were given a blanket order to 'maintain the reasonable degree of order within the prison necessary for its effective functioning'.

The next day, the Stanford University campus police detained the 11 other students on suspicion of burglary. The police screeched up in front of their houses with sirens wailing and clapped them in handcuffs as the neighbours peered out of their windows to see what was going on. They were taken blindfolded to the prison, where they were required to strip. They were then photographed, treated with a delousing powder and issued with their prison clothing, which consisted of a kind of white smock with a number on the front and back, with no underwear allowed underneath. They also had to wear plastic sandals and a nylon stocking as a cap, and a chain with a padlock was slung around one of their ankles.

During their short period of simulated incarceration, Zimbardo tried to inculcate in his prisoners the same kinds of emotion that genuine detainees experience after a long spell behind bars: powerlessness, dependence and hopelessness. The clothes were designed to humiliate the prisoners and to take away their individuality. The chain round their ankles served as a constant reminder of where they were, even in their sleep.

On the first day, the 16 rules that the guards had drawn up – with the help of David Jaffe – were read out: 'Rule Number One: Prisoners must remain silent during rest periods, after lights are out, during meals and whenever they are outside the prison yard. Two: Prisoners must eat at mealtimes and only at mealtimes. ... Seven: Prisoners must address each other by their ID number only. ... Sixteen: Failure to obey any of the above rules may result in punishment.' Several times during each shift – even in the middle of the night – the warders had the power to muster the prisoners for a roll-call, where they were required to state their identification number and recite the 16 prison rules. At the beginning, these inspections took ten minutes; later, they could last for up to an hour.

Intriguingly, Zimbardo didn't really have any hypothesis about what might happen in such a situation. The somewhat vaguely formulated aim of the experiment was to find out what the psychological effects of being a prisoner or a prison guard might be. He wanted to understand exactly how prisoners lose their freedom, independence and privacy, while the warders become more powerful by controlling the prisoners' lives. His earlier experiments had shown how easily quite normal people can be induced to commit inhuman acts, either by placing them in a group where they were no longer perceived as individuals or by putting them in a situation where they regarded other people as enemies or objects. The 'Stanford prison experiment', as it is now known, combined several of these different mechanisms. It became so well-known that a rock group in Los Angeles even named itself after it.

On the second day – after a roll-call at 2.30 in the morning – the prisoners rebelled. They took off their caps, ripped the numbers off their clothes and barricaded themselves in their cells. The guards used a fire extinguisher to drive them away from the door and then meted out punishment: the ringleaders were shut in the 'hole', a dark box at the end of the corridor. Those who hadn't

From very early on in the experiment, the guards used the roll-call as an opportunity to bully the prisoners.

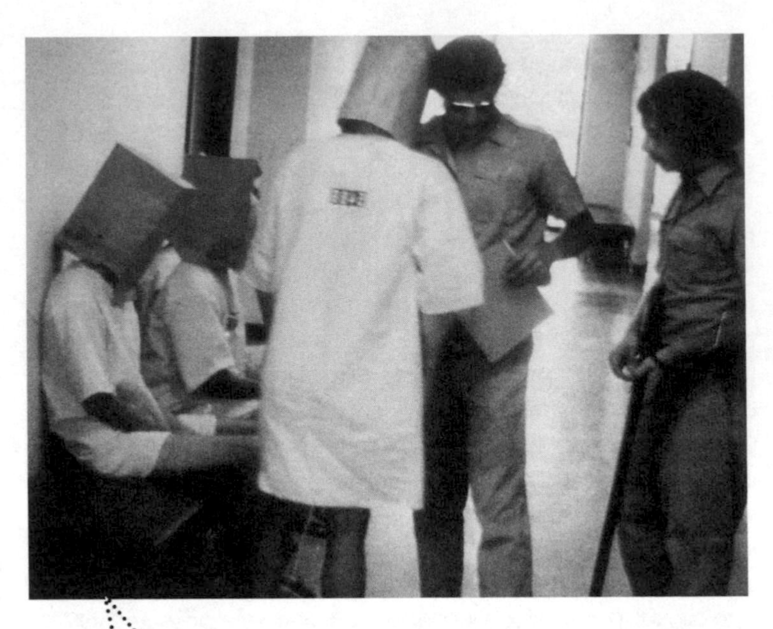

Prisoners had to make their regular evening visit to the toilet wearing paper bags over their heads.

played any part in the riot were given preferential treatment in a special cell and received better food. Shortly thereafter, without any warning or explanation, the guards put people from these two groups into the same cell. This disoriented the inmates and made them begin to mistrust one another. They never again rebelled as a group.

The warders then proceeded to institute absurd rules, discipline the prisoners arbitrarily and give them pointless tasks to do. For instance, they were made to move crates from one room to the next and then back again, clean the toilet with bare hands or pick thorns out of their blankets for hours on end (after the warders had dragged the blankets through briars). They were also ordered to mock their fellow inmates or to simulate sex acts with them.

madsciencebook.com Original footage of the Stanford Experiment.

After less than 36 hours, Zimbardo was obliged to release prisoner 8612, who was suffering severe depression, uncontrollable bouts of sobbing and fits of rage. He hesitated at first, thinking that the student was just pretending to be at the end of his tether. Zimbardo found it inconceivable that a participant should exhibit extreme reactions of this kind after such a short spell in a mock prison. However, over the next three days the same thing happened with three other subjects.

A misunderstanding had resulted in the test subjects believing that they had no option of quitting the experiment voluntarily.

Gradually, the boundaries between experiment and reality began to blur, for both the prisoners and the guards. The longer the experiment went on, the more often the guards had to be reminded that no physical violence was allowed. The power that the experiment gave them turned students who'd been inclined towards pacifism into sadistic prison warders. Even Zimbardo began to behave oddly. One day, one of the guards thought he had overheard the prisoners planning a mass breakout. 'How do you think we reacted to this rumour?' Zibardo later wrote. 'Do you think we recorded the pattern of rumour transmission and prepared to observe the impending escape? That was what we should have done, of course, if we were acting like experimental social psychologists. Instead, we reacted with concern over the security of our prison.' What Zimbardo actually did was to go to the Palo Alto police to try to get the prisoners transferred to the old city jail. When the police refused, he flew off the handle and started to yell about the lack of cooperation between prisons. Zimbardo had himself become a prison superintendent. In any event, the planned breakout never happened.

madsciencebook.com **Original materials from the experiment (e.g. the 16 prison rules and the outline of the experiment that was handed to the participants before it began), along with studies on it and reactions to it are available at www.prisonexp.org.**

The next thing Zimbardo feared was that after visiting hours, the students' parents would insist on taking their sons home. So he ordered the prison to be made spick and span, while the prisoners were given good food and permitted to wash and shave. The visitors were greeted by a pretty young woman. They had to report to reception, wait for half an hour and were then granted ten minutes' visiting time. Some of the parents were shocked at the prisoners' haggard state, but even they seemed to take the prison at face value and ended up asking the superintendent individually if he could provide better conditions for their sons.

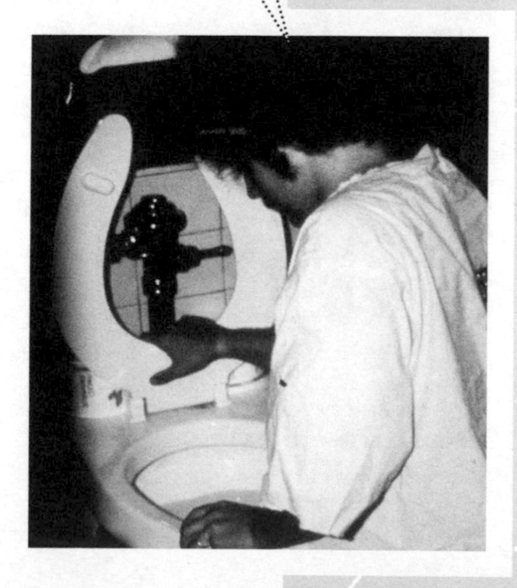

The warders imposed humiliating punishments such as forcing prisoners to clean the toilets with their bare hands.

A short time later, Zimbardo summoned a Catholic priest who had worked in jails before. Half the prisoners were asked to line up and repeat their numbers to him. Unbidden, he also slipped straight into the role of a real prison chaplain. Although the prisoners hadn't committed any crime and Zimbardo had absolutely no legal powers over them, the chaplain advised them to seek the help of a lawyer to secure their release.

The Stanford prison experiment made such a powerful impression on a group of Los Angels musicians that they named their band after it.

On the fourth day, Zimbardo formed a parole board from secretaries and doctoral candidates from the Psychology Faculty, to which the prisoners could appeal for early release. Almost all of them were prepared to waive the $15 a day they were due in exchange for being let out. This 'Adult Authority Parole Board' sent them back to their cells while they deliberated over their pleas. Amazingly, all the prisoners complied, even though they could have curtailed their participation in the experiment without more ado by just refusing to accept the payments anyway. But they didn't have the stamina to do so. 'Their sense of reality had shifted, and they no longer perceived their imprisonment as an experiment. In the psychological prison we had created, only the correctional staff had the power to grant paroles.'

In the meantime, a lawyer showed up, whom the parents had contacted to get their sons released. He spoke with the prisoners to ascertain how they planned to post bail and promised to return after the weekend – although even he knew that this was just an experiment and the question of bail was ludicrous. At this point, everyone involved had completely lost sight of where their role ended and their true identity began.

For the duration of the experiment, the psychologist Philip Zimbardo played the role of the 'prison superintendent'.

On Thursday evening, five days after the start of the experiment, Zimbardo's girlfriend and later wife Christina Maslach visited the jail. She was a psychologist herself, and had agreed to interview the prisoners the next day. There wasn't much going on at the time, so Maslach sat down in the control room to read an article. At around 11 p.m. Zimbardo tapped her on the shoulder and pointed at the closed-circuit TV screen, shouting 'Quick, quick – look at what's happening now!' Maslach looked up and felt instantly sick. The warders were screaming at a row of prisoners chained together at their ankles, and whose heads were concealed in paper bags. This was the nightly toilet run before lights out. If the prisoners were caught short in the night, they had to relieve themselves in their cells in a bucket, which the guards randomly refused to slop out. 'Do you see that? Come on, look – it's amazing stuff!' Zimbardo went on. But Maslach really didn't want to look. When Zimbardo asked her what she thought of the experiment as she was leaving the jail, she replied, 'What you are doing to those boys is a terrible thing!' A massive argument ensued, in the course of which Zimbardo realized that everyone involved with the experiment had internalized the destructive aspects of prison life. When it was over, he decided to call a halt to the experiment the next morning.

Source

Haney, C., et al. (1973), 'Interpersonal Dynamics in a Simulated Prison'. *International Journal of Criminology & Penology* 1 (1), pp. 69–97.

Zimbardo, P.G. (2007), *The Lucifer Effect*. Random House.

Because Zimbardo had his prison courtyard watched round the clock by a video camera, his study is regarded as a precursor of the now-familiar reality-TV format – with the important distinction that it wasn't being conducted with an eye to securing big ratings. But even that wasn't long in coming. In 2002 the BBC launched a reality show with the title *The Experiment*, which was designed to reprise the Stanford experiment in front of millions of TV viewers. Zimbardo regards the results of this experiment, at which two psychologists were present, as highly questionable, since the participants were aware the whole time that they were being filmed.

A follow-up study made one year after Zimbardo's experiment revealed no lasting detrimental effects on any of the participants. Prisoner 8612, who had been the first to crack, later became a psychologist at the county jail in San Francisco.

The most important outcome of the Stanford experiment was the realization as to how powerful the force of circumstances can be. As in the Milgram experiment (see p. 167), perfectly normal people exhibited completely unexpected behaviour in an unfamiliar situation. Clearly, an individual's personality could not be used to predict his behaviour if he finds himself in a situation where he knows none of the ground rules. As Zimbardo wrote after the experiment, 'any deed that any human being has ever done, however horrible, is possible for any of us to do – under the right or wrong situational pressures ... That knowledge does not excuse evil, rather, it democratizes it, shares its blame among ordinary participants, rather than demonizes it.'

It's hard to accept this unpalatable fact about human nature. After all, who wants to believe that they have it in them to be a torturer? To try to bring some clarity to what had happened at Abu Ghraib prison, Philip Zimbardo appeared as an expert witness in the trial of one of the accused soldiers, 37-year-old Chip Frederick. However, his testimony had scarcely any effect on the outcome. Frederick was sentenced to eight years' imprisonment. Zimbardo commented: 'They had to show the world and the Iraqi people that they were "tough on crime" and would swiftly punish these few rogue soldiers, the "bad apples" in the otherwise good US Army barrel.' Zimbardo wasn't trying to excuse Frederick's actions, but still found the sentence too harsh. It failed to take into account the fact that the army top brass was negligent in making soldiers into prison warders without training them for the role, and without issuing them with any specific orders, in circumstances in which they were subject to a high level of stress. Even so, Zimbardo's famous experiment wasn't totally without influence in the case of Abu Ghraib. While flying from Hawaii to Baghdad in May 2004, Colonel Larry James watched the documentary film *Quiet Rage* about the Stanford experiment over and over again. A military psychologist, James had been charged with the task of reimposing order and discipline at Abu Ghraib. He even invited Zimbardo to accompany him to Baghdad, but he was 'too fearful', as he later admitted. Ultimately, James drafted a series of rules to be implemented at the military prison, after which no more instances of abuse took place there.

As regards the remarkable story of the Stanford experiment, what's surprising is that the movie industry only cottoned on to the subject matter some thirty years after the event. A German feature film entitled *Das Experiment* ('The Experiment') was screened in March 2001. In it, a psychologist puts twenty students in a mock jail, ten in the role of prisoners and the other ten as guards. But after three days, things get out of control. The guards start tying up and beating the prisoners. Rapes and murders ensue. In the opening credits, the film makers had informed the audience that their work was based on the Stanford prison experiment. This led to Zimbardo receiving hundreds of e-mails from Germany. 'How on earth could you do that,' asked his correspondents. In the end, Zimbardo engaged the services of an attorney and succeeded in getting the reference to the experiment withdrawn. However, he failed in his attempt to stop the film being distributed in the USA. There are now plans afoot to make a new film of the story, based on Zimbardo's own account in his book *The Lucifer Effect*, and with Oscar-winner Christopher McQuarrie as director.

1971 Tip for Hitchhikers 2: Be Female!

The results of their study wouldn't 'depart radically from general expectations', wrote Margaret M. Clifford and Paul Cleary in their manuscript 'The Odds in Hitchhiking'. After a number of men and women, variously attired, had stood at the roadside for hours at a time, two key findings emerged: that it was harder to get a lift if you were shabbily dressed, and that women stood a better chance of being picked up than men. Clifford and Cleary also assessed the effect of group size and composition on the behaviour of car drivers: two men were the worst combination, while one person on their own got as many offers of a lift as a male-and-female couple. But two women together had the best chance of all. Clifford and Cleary didn't bother evaluating the experimental condition of 'a woman on her own'. 'Our two females hitchhiking in sports attire appeared to be attracting such an inordinately large number of rides that a state trooper warned the girls about "disrupting traffic".' The next tip for hitchhikers can be found on p. 240.

Source

Clifford, M.M., and Cleary, P. (1971), 'The Odds on Hitchhiking.' Unpublished manuscript.

1971 Galileo on the Moon

Despite the fact that Galileo Galilei had already performed an elegant experiment in the 17th century (see p. 14) to demonstrate that the velocity of a falling object has nothing to do with its mass, we still find this hard to accept. The trouble is our daily lives constantly furnish us with experiences that seem to indicate the opposite: thus, a bottle falls faster than a leaf, a hailstone faster than

Source

Allen, J. (1972), 'Apollo 15 Preliminary Science Report',Chapter 2: 'Summary of Scientific Results', (SP-289). NASA, p. 11.

a snowflake, and a hammer faster than a feather. Of course, our school physics teacher dinned it into us that these different fall velocities have to do with air resistance and not with the objects' mass. Nevertheless, the evidence of our own eyes continues to exert a powerful influence on us.

Watch astronaut David Scott simultaneously dropping a hammer and feather onto the Moon's surface.

madsciencebook.com

And so, on 2 August 1971, the astronaut David Scott conducted an experiment that was captured on film. Standing on the Moon – which has no atmosphere – he simultaneously dropped a feather and a hammer weighing 40 times as much. Both hit the surface of the Moon at exactly the same moment. Although the result was a foregone conclusion, it was still reassuring, as a NASA report on the Apollo 15 mission later stated. After all, the astronauts' ability to get home safely depended entirely on the theory associated with this experiment holding water.

1971 The Atomic Clock Flies Economy

'The thing weighed a ton,' Joseph Hafele vividly recalls. 'When we lugged it onto the plane, it sat at an angle between us, with me bearing most of the weight.'

Together, the two atomic clocks that Richard Keating from the US Naval Observatory and Joseph Hafele from the University of Washington strapped on board the Boeing 747 at 7.30 on the evening of 4 October 1971 weighed 132 pounds (60 kg). The size of a chest of drawers, they needed two seats of their own – and consequently their own tickets, issued in the name of 'Mr Clock'.

'The clock's ticket was $200 cheaper than ours – but then it wasn't eating during the flight,' explains Hafele, who in the course of the journey gave up using the term 'atomic clock': 'During a stopover in Istanbul, journalists kept asking me what our experiment had to do with the atomic bomb.' In addition, their fellow passengers recoiled in horror when the scientists explained to them in all innocence that the box on the seat next to them was an atomic clock. 'After that, we took to calling it a caesium clock.'

Another passenger glanced at the hand on the caesium clock, then at his own wristwatch, and announced, 'Your clock's a bit fast.' The man couldn't have known that what was facing him was one of the most accurate clocks in the world. It was only

The physicists Joseph Hafele (left) and Richard Keating with the atomic clocks that they twice took round the world.

through the use of such a clock that a strange prediction made by a technical expert (third class) from the Berne Patent Office in 1905 could be put to the test.

In was in that year that Albert Einstein, then 26 years old, turned the world of physics on its head with his publication of a two-page article. Under the unassuming title 'On the Electrodynamics of Moving Bodies', Einstein formulated for the first time the theory that was later, under its familiar designation of the special theory of relativity, to completely reshape our view of the world. Among other radical proposals in this paper, Einstein dispensed entirely with the concept of absolute time. Time doesn't pass equally quickly everywhere, he claimed, but rather is dependent upon speed. Time passes more slowly for someone travelling faster. In asserting this, Einstein wasn't thinking of individuals' perception of time but of time as a physical dimension. For anyone moving at a greater speed, clocks ticked more slowly, water took longer to boil and a game of chess lasted longer. The individual in question wouldn't notice any of this, because his perception of time would also be slowed down – in other words, he would age less quickly. A twin who took a flight in a rocket would notice the effect when he met his brother again on his return, for, despite being born on the same day, the latter would now suddenly be older.

This proposition seemed absurd to both laymen and experts. Einstein's theory wasn't borne out by any experience in everyday life. And no wonder. To measure the effect, one would either have to travel at enormous speed, close to the speed of light, at some 186,000 miles per second (300,000 km/sec), or to have an incredibly accurate timepiece.

Ten years later, Einstein formulated his general theory of relativity, which among other things states that the motion of a clock is determined not merely by the speed it is travelling at but also by gravity. A clock runs faster on a mountain than in a valley. Once again, the difference on Earth is so infinitesimal that we don't notice it.

Hafele and Keating getting off the plane. The clocks weighed 132 pounds (60 kg).

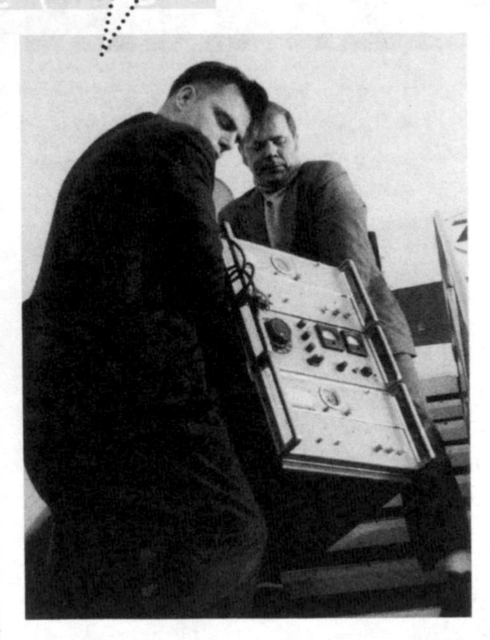

After US airlines inaugurated round-the-world flights at the beginning of the 1970s, Joseph Hafele wondered whether it might not simply be possible to take a clock on board to measure the effect Einstein predicted. It was all quite elementary, really. Before the flight all you'd have to do would be to synchronize the travelling clock with one that was going to remain stationary on Earth, and then fly round the world. If Einstein was right, at the end of the process, a time differential would become apparent, with the clock that had moved faster running slow.

Hafele calculated that the time differential should be a few billionths of a second. When he delivered a lecture presenting his ideas at a physicists' congress, Richard Keating of the Time Service Division of the US Naval Observatory in Washington was in the audience. At that time, this arm of the US military was responsible for keeping the exact time for the 'free world'. Knowing the precise time was especially vital in radio navigation, Keating had often carried a portable atomic clock with him on flights abroad to synchronize atomic clocks at other US facilities around the globe.

Keating realized immediately that such a clock would be

capable of measuring the time differential that Hafele had calculated, and so the two got together and planned their trip round the world. Yet Keating himself was sceptical as to whether any difference would be noticeable: 'I don't trust these professors who scribble something down on a blackboard and claim to know everything. I've made too many measurements in my time that didn't turn out as predicted.'

Hafele and Keating first flew round the world from west to east and then, four days later, repeated their journey in the opposite direction. Their first journey, lasting 65 hours in total, took them from Washington DC to London and from there to Frankfurt, Istanbul, Beirut, Tehran, Delhi, Bangkok, Hong Kong, Tokyo, Honolulu, Los Angeles, Dallas and finally back to Washington.

These flights were an arduous business. Not just because the physicists had to repeatedly hump the heavy clock around (in fact, it was two clocks put together, which made the measurements more accurate). They hardly got an opportunity to sleep, as the fragile instruments required constant monitoring. And if all this wasn't enough, because of faulty wiring in the apparatus, Keating was forced to disconnect the earth, which meant that the clock's casing became live and regularly gave them electric shocks.

At the end of the flights, Hafele fed the data they had collected into Einstein's equation and, after factoring in the effect of speed and gravitational pull, calculated that the clock in the plane should be running between 17 and 63 billionths of a second slower. And indeed, in actual fact it was running 59 billionths of a second slower. Because the clock that had stayed behind in Washington was turning with the Earth, the westward journey yielded the opposite result. Now, the time in Washington was passing more slowly than on board the aircraft, to the tune of 273 billionths of a second. At first, this seemed odd; after all, in this case as well, the clock in the plane was still going faster than the one on the ground. But the thing looked different when viewed from outside: in rotating with the Earth, the clock on the ground was actually going faster than the one in the plane, which was flying in the opposite direction to the Earth's rotation.

To the annoyance of some physicists, who regarded the experiment as unnecessary, the flight of the atomic clock got a lot of media coverage. For sure, the special theory of relativity had been proved several times already in experiments, the first of which had taken place in 1938. But these early tests measured the difference in the time it took for elementary particles that

madsciencebook.com **An animation showing the effect of the special theory of relativity on time.**

Source

Hafele, J.C., and Keating, R.E. (1972), 'Around-the-World Atomic Clocks: Predicted Relativistic Time Gains'. Science 177, pp. 166–168.

had been accelerated to high velocities to decay. Such experiments weren't all that intelligible to the layman. The clock that Keating and Hafele carried on board with their hand luggage was far more real. As Hafele said, 'At some stage I decided that I did this experiment for the average guy, not for experts.'

The celebrated physicist Stephen Hawking wrote, 'If you wanted to live longer, you could keep flying to the east so the speed of the plane added to the Earth's rotation.' But at the same time, he warned people not to raise their hopes: 'However, the tiny fraction of a second you gained would be more than offset by eating airline meals.'

1972 A Quick Getaway at the Lights

How do you begin to study the effect on people of being stared at? Researchers from Stanford University developed a simple method – they stood at a number of crossroads in Palo Alto and stared at car drivers waiting at the traffic lights. As soon as the lights turned to green, they measured how quickly the cars drove off. The drivers who had been stared at were keen to get away, crossing the junction in 5.5 seconds on average, while other drivers took 6.7 seconds. Just like animals, human beings evidently interpret being stared at as a threat, causing them to take flight – even if in this case the 'flight' only took them across a road junction.

Source

Ellsworth, P.C., et al. (1972), 'The Stare as a Stimulus to Flight in Human Subjects: A Series of Field Experiments'. *Journal of Personality and Social Psychology* 21, pp. 302–311.

1973 The Sex Raft

This was an experiment that was just made for the tabloid press: 'Skippered by a buxom Swedish blonde, a raft carrying a group of bare-chested men and bikini-clad women who experimented for 100 days with "group and sexual behaviour" ended its 5000-mile Atlantic odyssey on Monday.' Thus ran a communiqué from the UPI press agency, reporting the arrival of the vessel Acali at the Mexican port of Cozumel on 20 August

1973. This marked the end of the 'largest group experiment in modern behavioural research'. At least, that's what the Mexican anthropologist Santiago Genovés called his brainwave of crossing the Atlantic on a living-room-sized raft in the company of six women and five men of various races and religions. Most newspaper readers, on the other hand, knew the Acali simply as 'the sex raft'.

In 1969 and 1970 respectively, Genovés had been a crewmember on *Ra I* and *Ra II*, the craft constructed out of papyrus reeds by the Norwegian anthropologist Thor Heyerdahl. As with his *Kon-Tiki*, Heyerdahl wanted to use these expeditions to corroborate his theories about epic voyages made by early peoples. During these voyages, Genovés learned what every sailor has known since time immemorial: 'There's no better test-bed for studying human behaviour than floating around together on the high seas in a cockle-shell.'

The raft that Genovés had built was 40 ft (12 m) long by 24 ft (7 m) wide. The cabin – a single room in which everyone slept – measured 14 ft (4 m) square, and was only chest-high. As a boat, the Acali hardly cut a dash, but rather resembled a floating island that drifted lazily along. On this craft Genovés assembled 11 guinea pigs: the Swedish captain, a Jewish woman doctor, a male photographer from Japan, a Greek restaurateur, an Angolan priest, a white American woman and an African American woman, an Arab woman from Algeria, a Uruguayan man, a Frenchman and himself.

Genovés hadn't had harmonious coexistence in mind when he selected this group; quite the contrary – this mix of people was chosen to be as explosive as possible. He deliberately appointed women to the two most important roles on board: skipper and ship's doctor. He was careful to ensure that as many participants as possible were married and had children, and that a wide range of races and religions were represented.

On 13 May 1973 the Acali set sail from Las Palmas in the Canary Islands. Shortly beforehand, Genovés had made known the sleeping arrangements in the poky cabin: there would be two lines of bunks, with men and women alternating. The others immediately accused him of positioning himself between the two most attractive women. Later, he himself complained that all people were interested in were the sexual aspects of the experiment.

During the 101-day crossing, Genovés filled over a thousand

Six women and five men of different races and religions crossed the Atlantic on the raft *Acali*.

pages with his observations of life on board. In addition, the participants filled out 46 questionnaires, in which they gave a total of 8079 answers on such diverse topics as relations on board, their sexual behaviour, religion, aggression and questions of morality.

At first, people were a bit reticent. Nobody wanted to reveal any weaknesses. The quickest inhibition to be dropped was people's reluctance to use the open-air toilet in full view of their fellow crewmembers. After a fortnight, it was possible to chat with anyone while they were doing their business. The first cause of friction on board was provided by the work rota. Captain Ingrid's hectoring tone soon began to grate on people's nerves. Aischa, the Algerian, shirked her duties and earned herself the nickname 'the tourist'. Almost everyone was riled by the Frenchwoman Sofia's excessive primping and preening, as she took over an hour to get ready every morning. Meanwhile, the priest exuded a constant and almost unbearable stench of BO. Genovés tipped him the wink, after which he washed himself from head to foot three times a day.

After two weeks Genovés asked himself how much sex was taking place on the 'sex raft'. And immediately answered his own question: 'Not much.' One of the six reasons he advanced for this state of affairs ran as follows: 'Some of them are still suffering from frequent bouts of seasickness and are throwing up all the time. Not very alluring.' The Japanese photographer Komico and the American Ana had clearly struck up a relationship. Genovés thought he had seen them making love by moonlight in the cabin. He himself became intimate with Sofia. After a month, Genovés recorded in his diary that a 'liberal and healthy but ultimately rather empty sense of camaraderie' had developed on board.

Questionnaire number 5 caused something of a stir. It contained such questions as: 'What disturbs you most about life on board the *Acali*?'; 'What do you like best about you and your fellow crewmembers?'; 'What do you like least?'; 'Would you like to see the bunks rearranged in the cabin?'; 'If so, who would you like to sleep next to?'; 'Who wouldn't you like to sleep next to?'; and 'If there were no inhibitions, whom would you most like to sleep with?' Everyone was agog to learn the answers. After this, new sleeping arrangements were agreed.

The crew's embarrassment at using the open-air toilet soon evaporated.

On 13 July one of the *Acali's* rudder blades sheared off. Despite the fact that the sea was full of sharks at the time, Genovés jumped in to inspect the damage. Suddenly, everyone knew exactly what they had to do. 'Do life-threatening situations have to arise, then, before there's any team spirit among the crew?' Genovés asked himself. As it turned out, the rudder could be replaced.

After seven weeks, Ana proposed playing the truth game: everyone asked a person of their choice four questions in writing, which were then read out anonymously and answered in front of the whole group. For example, Genovés was asked, 'When you're on one of your expeditions, does your wife have extramarital affairs as well, then?' His response was, 'I don't think so, but I don't know for sure.' Emiliano was asked, 'Would you like to sleep with a woman?' He replied, 'If somebody really liked me, then I wouldn't say no.' And Antonio was challenged, 'How can anybody be as two-faced as you are?' to which he replied, 'I don't think I am two-faced.'

After two months, Genovés tried using shocking questions to find out how the participants would react to a deliberate infringement against convention: 'Should we … spend a whole day naked?' The result was six in favour and five against. 'Or hold a kind of party at which everyone sleeps with everyone else?' – four in favour, the rest against. 'Should we ban people from forming couples?' – two for, the rest against. And finally, 'Should we maintain the status quo?' – two for, six against, three abstentions.

Women Will Boss Men in Sex Study

The press often reported that women occupied all the important posts on board.

When 13 weeks had passed, the two American women suggested that for a period of five nights, a man and a woman should be left alone in the cabin together for an hour at a time. The suggestion was turned down, but Genovés recognized the need for people to have some time to themselves now and again and so proposed that five randomly drawn couples should be given the opportunity of getting together at the five places on the raft that couldn't be seen by anyone else. A mood of great anticipation spread through the boat. After these assignations, there was a great deal of vulgar banter in the galley, as dirty jokes and insinuating remarks were bandied about. Genovés confided to his diary: 'I'm a bit depressed. The intellectual tone on the raft is definitely getting lower.'

Thereafter, events came thick and fast. The Japanese photographer Komico tried to jump overboard. He hated the pictures he took, found it hard getting on with the others and was spurned by his lover Aischa. At around the same time, a freighter almost rammed the *Acali*, and Genovés got appendicitis. As in earlier crisis situations, this concentrated the group's minds and made them get their act together. Genovés's appendicitis improved. Two weeks later, the *Acali* entered Cozumel, where all the subjects were immediately isolated and watched over by armed security guards in their hotel. For a whole week, they were subjected to a battery of tests by psychiatrists, psychologists and physicians.

These follow-up studies didn't yield much in the way of results. And, despite Genovés's protestations to the contrary, neither did the whole experiment. In his book *Acali*, which was published in 1975, he interprets everything that happened on board in such a way that it fits in with his own worldview – which is that of a person steeped in high culture. On board the raft, he believed he had found the 'new man', who was 'free from all fateful territorial ambitions and all aggressive or sadistic impulses'. Where sexuality was concerned, Genovés came to the following conclusion: 'There is no innate sex drive that can sufficiently explain the apparently irresistible urge we have to form sexual relationships.'

A group meeting on the cabin roof. Conflicts erupted time and again between those taking part in the experiment.

Genovés's experiment came in for some swingeing criticism. Even during the voyage, his colleagues at the university had distanced themselves from the affair. Many found it unethical that he had got all the participants to sign an undertaking beforehand entitling him to make use of the data that emerged from the voyage as he saw fit, 'even when it was of an intimate nature'. Granted, in his book he didn't use people's real names, but then again it is copiously illustrated and the participants are easily recognizable –and the participants' real names appeared in newspaper reports anyway.

This inconsistency also extended to Genovés's role on board the *Acali*. On the one hand, he didn't want to be group leader. He wasn't interested in finding out what would happen if this motley crew were forced to bury their differences and pull together. On the other hand, he berated the participants when they overslept or failed to fill in the diary that he'd asked them to keep.

Source

Genovés, S. (1975), *Acali*. Planeta.

However, all this criticism was like water off a duck's back to Genovés. He regarded his experiment as a major contribution to human coexistence. Twenty-five years later, canny television producers came up with reality shows like the pioneering Swedish series *Expedition: Robinson* and *Big Brother*, which were the spitting image of the *Acali* experiment. Genovés could have seen this coming, given that the *Acali* was partly funded by a Mexican TV station.

1973 Trembling Knees Make Hearts Flutter

Whenever Japanese researchers are visiting the psychologist Donald G. Dutton, he always has to take them to see the Capilano suspension bridge. This bridge is a local tourist attraction near Vancouver: 5 ft (1.5 m) wide, almost 500 ft (150 m) long, and made of ropes and rickety wooden boards, it spans a 230-ft (70-m)-deep gorge. But this isn't what draws the Japanese psychologists there. They are keen to see it because it was here that Dutton and his colleague Arthur P. Aron carried out a famous experiment to investigate love's tortuous path.

Here's how it went. In the summer of 1973, visitors to the suspension bridge were importuned by a pretty female student, whom Dutton and Aron had hired to spin a yarn to any man who'd just stepped off the bridge with his knees a-tremble. She

The Capilano rope suspension bridge near Vancouver was the site of a famous experiment on the aberrations of love.

told them that she was doing an assignment for her psychology degree on the effect of scenic attractions and asked if they'd mind answering a few questions. Unbeknownst to the men, what happened then was the key moment of the experiment: the student tore off a corner of the questionnaire, jotted down her name – Gloria – and telephone number on it, gave it to the men and asked them to give her a call later if they wanted to know more about the survey.

Sometime later, this same student was to be found hanging about a small park near the bridge. Men who had crossed the bridge a while earlier were taking a breather here. She approached several of them, telling them the same story and giving them her phone number. But there was one small difference: now she was no longer called Gloria, but Donna.

Over the following days, 13 of the 25 men who'd been approached at the end of the bridge called Gloria, while 7 of the 23 men from the park wanted to speak to Donna – just as Dutton and Aron had predicted. The suspension bridge provided further proof of a hypothesis that had long been the subject of discussion in psychology: if a particular stimulus engenders physical excitement in people, is it possible for them to then falsely attribute this excitement to another stimulus? Psychologists refer to this as 'misattribution'.

The men who'd just stepped off the bridge misinterpreted Gloria as the cause of their trembling knees. Subconsciously, they thought that the student was responsible for their physical arousal, which in actual fact had been caused by crossing the rope bridge. Little wonder, then, that so many of them wanted to get in touch with Gloria afterwards. As for the men in the park, their excitement at crossing the bridge had already abated, and as a result there were no physical signals present for which Donna might have been misread as the cause.

Since this experiment, misattributions have been shown to operate in a number of different contexts. Some researchers use the phenomenon as the basis of their hypothesis that parents who forbid a young couple from seeing one another only succeed in creating additional arousal, which is then misinterpreted as feelings of love, thereby further strengthening the bond between the young people. For these psychologists, then, the story of *Romeo and Juliet* is a classic case of misattribution.

Source

Dutton, D.G., and Aron, A.P. (1974), 'Some Evidence for Heightened Sexual Attraction under Conditions of High Anxiety'. *Journal of Personality and Social Psychology* 30, pp. 510–517.

1973 Spiders 4: in Space

By the 1950s scientists had already found out how a spider builds its web when under the influence of drugs (see p. 123), with some of its legs missing (see p. 136) or after having drunk the urine of schizophrenics (see p. 139). With the dawning of the space age, the question naturally arose as to what a web spun under conditions of weightlessness would look like. As early as 1968 the biologist and spider specialist Peter N. Witt drew up plans to send spiders into space on board a satellite. But it was only four years later – when a schoolgirl, Judith Miles, suggested a spider experiment in space in a competition sponsored by NASA – that it actually came to fruition.

On 28 July 1973 an Apollo rocket with three astronauts and two spiders – Arabella and Anita – on board blasted off, bound for the US space station Skylab. There Arabella was placed in a frame, in the corners of which she wove a couple of pathetic webs the next day. But shortly thereafter, she clearly became accustomed to weightlessness and constructed a complete web, as did Anita just a bit later. However, they sat and waited in vain for prey to land in their webs: no flies had been brought on the journey. If garden spiders (Araneus diadematus) get sufficient water, they can survive for up to three weeks without food. Twice, the astronauts departed from the script and hung a tiny piece of filet mignon steak in the webs, which both Arabella and Anita took immediately.

This is the frame in which the spiders wove their webs while in space. The camera captured the various stages of construction.

NASA handed over photographs of the webs built in space to Witt for him to evaluate, but he was clearly less than impressed with what they'd supplied. Time and again in his paper on the subject, he complains about the poor quality of the photos, which he said made a proper analysis very difficult. In addition, he claimed that NASA hadn't kept an accurate record of which webs the various samples of spiders' silk that were brought back from the mission came from.

According to Witt, the most important finding to emerge from the test was that spiders can adapt to weightlessness and can build a perfectly passable flytrap even under these unfamiliar conditions. Web construction occupies a high priority among spiders, since no web means no food.

The story of Arabella and Anita's spaceflight has a sad ending. At the end of their stay on Skylab, they joined the ranks of fallen space heroes and died, as scientists later discovered, of dehydration.

Animal lovers, though, can take solace from the fact that the spiders did not die in vain: looking at the space webs they had woven, one of the NASA research team identified a new method of constructing tennis rackets, which he planned to market under the name 'Rocket Racquet'.

Source

Witt, P.N., et al. (1977), 'Spider Web-Building in Outer Space: Evaluation of Records from the Skylab Spider Experiment'. *American Journal of Arachnology* 4, pp. 115–124.

1973 Invasion in the Urinal

Dennis Middlemist got the idea for this experiment when he was sitting on the toilet, and that was also precisely the place where he planned to carry it out. At the time, Middlemist was attending a seminar on environmental psychology and was mulling over an assignment the class had been set. One topic in particular fascinated him: personal space. How much space does a person need around them? Why is such a thing necessary? And what happens if someone invades it?

It was an everyday occurrence that provided him with initial answers to his questions. 'I immediately noticed the effect one day when I was standing at the urinal and one of my fellow students was using the one next to me.' What Middlemist noticed was that it took longer for him to relieve himself. But when he suggested to the seminar that they conduct an experiment on personal space based on this observation, his classmates laughed him out of court. Despite this, the professor encouraged him to give it a go anyway.

There was one big problem with investigating the phenomenon of personal space. Sure, it was clear from experiments that people reacted when the invisible boundaries around them were violated. For instance, they recoil to try to restore the space or they try to compensate for excessive proximity in a variety of other ways. But what nobody knew was why. This is where Middlemist's urinal experience came in.

It had long been known that feelings of fear and apprehension affect how quickly a person's sphincter relaxes. Basically, anybody who's anxious or upset takes longer to go. So, if an unsuspecting test subject was to discover that his peeing was delayed by the presence of another man, this would provide an elegant proof of the notion that violating personal space engenders feelings of anxiety and apprehension.

To test this hypothesis, Middlemist and his fellow student Eric Knowles devised an experiment, which they ran towards the end of 1973 in a gents' toilet with three urinals, located near one of the large lecture auditoriums on the campus of the University of Wisconsin-Green Bay.

By using dummy 'out of order' signs, they engineered the following three scenarios for men using the toilet:

1. The test subject found himself standing directly next to one of the testers at the next urinal.
2. There was an empty urinal between the tester and the subject.
3. The test subject was alone in the toilet.

At least, in the last case, he thought he was alone. In actual fact, Middlemist would be sitting in one of the two cubicles behind the urinals with two stop-watches and a periscope concealed behind piles of books, which he could use to look out from under the cubicle door and observe the test subject's stream of urine. He started the watches going when the subjects stepped up to the urinal and stopped the first when they started to go and the second when they stopped.

Once 60 men had been and gone, the matter was clear. If a tester was standing at the next urinal, it took an average of 8.4 seconds for the subject's sphincter muscle to relax – almost twice as long, in other words, as when they were on their own. And, as predicted, a stressed piddler finished more quickly than a laid-back one, because a greater head of pressure had built up during the longer time they'd spent waiting for things to happen.

When Middlemist published these findings, he found himself accused of unethical conduct. Gerald P. Koocher of Harvard Medical School claimed that the experiment 'raises significant questions about the current state of human dignity as determined by psychological researchers', and went on to voice his concerns about 'unstable individuals who might accidentally discover that they were being observed during the course of the "micturition"…'. Middlemist found this all a bit over the top; after all, he reasoned, invasion of personal space was a daily occurrence in toilets. This wasn't an experience that did men any lasting harm. On the contrary, he suspected that men who found out in the pilot study that they had been part of an experiment wouldn't hesitate to make it the stuff of anecdotes to regale their friends with.

The urinal experiment dogged Middlemist for the rest of his life. As a result of it, he even found himself the focus of a political storm in a teacup. After he'd taken up a post at Oklahoma State University, the state governor got wind that some of his constituents were up in arms about the experiment. The governor complained to the college rector, who had to reassure him that it hadn't been conducted at his university.

Even if had been, the governor should have been happy: Middlemist's experiment was a conclusive and cost-effective bit of science conducted with the minimum of fuss. What more could a politician ask for?

Source

Middlemist, R.D., et al. (1976), 'Personal Space Invasions in the Lavatory: Suggestive Evidence for Arousal'. *Journal of Personality & Social Psychology* 33, pp. 541–546.

1974 In a Froth at the Lights

Car drivers get all worked up if, for no apparent reason, the car in front of them fails to move when the lights change to green. Not only is this a truism, since 1966 it even had the status of scientific fact (see p. 185). The psychologist Robert A. Baron wondered how a person's traffic-light rage might be mollified. In the course of various laboratory studies, he'd demonstrated that aggression diminished if a test subject was exposed to a stimulus that evoked other feelings, such as sympathy, amusement or sexual arousal. Now, Baron reckoned, it was time to road-test this hypothesis.

And so it came to pass that, in the summer of 1974, 120 car drivers in West Lafayette, Indiana, were treated to Baron's little psychodrama. At a set of traffic lights, a car driven by one of his accomplices blocked the vehicle behind him; there then appeared a well-endowed female student, dressed in a miniskirt and tight top, who crossed the street between the two cars. She was there to furnish the experimental condition of

Source

Baron, R.A. (1976), 'The Reduction of Human Aggression: a Field Study of the Influence of Incompatible Reactions'. *Journal of Applied Social Psychology* 6, pp. 260–274.

The experimental condition of 'sexual arousal': a full-breasted student in a miniskirt and tight top crossing the road.

'sexual arousal'. Less fortunate drivers encountered one of the other four experimental conditions: no female student (control group), a normally dressed female student (distraction), a female student on crutches (sympathy) and a female student with a clown's mask (amusement).

The accomplice's car blocked the road for 15 seconds at a time. The question was whether the various different conditions changed anything in the drivers' reactions. The hardly surprising outcome was as follows: if they caught sight of the crutches, the clown's mask or the miniskirt, the drivers left it longer before honking than they did if confronted with the normally dressed student. But ultimately the miniskirt was far more effective than either the clutches or the clown mask.

1974 Tip for Hitchhikers 3: Look them in the Eyes!

After science had succeeded in establishing two fundamental rules for hitchhikers – 'Get an injury' (see p. 187) and 'Be female' (see p. 223) – and proving them through experimentation, the year 1974 brought another breakthrough in how to influence car drivers. Out of 600 cars in Palo Alto, California, 40 stopped if the hitchhiker stared at the driver, but only 18 if there was no eye contact between the driver and the hitchhiker.

Source

Snyder, M., et al. (1974), 'Staring and Compliance: A Field Experiment on Hitchhiking'. *Journal of Applied Social Psychology* 4, pp. 165–170.

1974 Tip for Hitchhikers 4: Be Well-endowed!

And here's another handy and scientifically proven hint for female hitchhikers (sorry, chaps – this one only works for women). Try making your breasts

larger. In an experiment in Seattle, women wearing padded bras (i.e. ones that increased their breast size by more than 2 inches / 5 cm) got offered twice as many lifts as those without.

However, when Carol Fahrenbruch, one of those taking part, was later interviewed, she was quick to dispel any dreams of hordes of scantily clad women participants waiting by the roadside: 'Not to spoil the story a bit with an unfortunate fact and no doubt to the disappointment of your readers, the study was conducted during the cold and rainy Seattle fall and winter, so all of the "hitchhikers" were well wrapped in raincoats and ski jackets the whole time.' The official report on the study goes so far as to speculate that it might have been even more successful in less inclement weather: 'visibility of signals – particularly in the case of bust size – was probably hindered by frequency of rain and the necessity for heavier clothing worn by the hitchhiker'.

Source

Morgan, C., et al. (1975), 'Hitchhiking: Social Signals at a Distance'. *Bulletin of the Psychonomic Society* 5, pp. 459–461.

1975 Perspiration Extract in the Waiting Room

There was nothing remarkable about the waiting room at Birmingham University's dental clinic. It contained a reception desk, a low table strewn with magazines, and 12 chairs. The patients who sat down on them thought that they'd chosen one at random. But they were wrong.

Earlier that morning, Michael Kirk-Smith had gone into the waiting room, which was still empty at that time of the day, with an aerosol in one hand and a stopwatch in the other. He went over to the chair immediately opposite the reception desk, pointed the aerosol at the seat and sprayed for precisely five seconds. A fine mist containing 16 millionths of a gram of androstenone descended on the chair. He did the same thing every day over the next few weeks. Sometimes he'd spray for just a second and sometimes for ten. Often he'd wipe the seat beforehand with a cleaning agent and swap the chair with another one. Androstenone is a substance that men secrete in their armpits. Kirk-Smith was convinced that it attracted women.

Biologists had known for a long time that mating in many species of animal was regulated by volatile substances known

Source

Kirk-Smith, M., and Booth, M.A. (1980), 'Effects of Androstenone on Choice of Location in Others' Presence'. In Starre, H. van der, ed., *Olfaction and Taste* VII, pp. 397–400. IRL Press.

as pheromones. Similar substances had also been identified in human sweat. However, psychologists found the idea that they might play a role in people's choice of partners frankly ludicrous. 'They said that humans were too advanced for such primitive effects,' recalls Kirk-Smith, who was sure that they were a factor. He therefore began to study pheromones for his doctorate.

In the course of his study he got to know the psychiatrist Tom Clark. Clark had already conducted a few small tests of his own with androstenone. 'He had sprayed androstenone onto a seat during a party and told me that "Only homosexual men sat on it." I asked him how he knew they were homosexual. He replied along the lines "I'm a psychiatrist – I just know these things." Clark also ran an experiment in a theatre auditorium, spraying programmes and certain seats. However, there was no real scientific method to his tests. Kirk-Smith's experiment, on the other hand, was to be rigorously designed and conducted.

For the first four days of the test, he just watched, without spraying any androstenone. An assistant in the dental practice, who wasn't told the purpose of the experiment, kept a log throughout the day of which seats men sat on and which ones women went for. The three chairs opposite the reception desk were clearly unpopular among women. For instance, not one of the 67 female patients who visited the clinic during this period sat down on the middle one. This was the chair that Kirk-Smith intended to use to demonstrate the power of pheromones.

Over the next five weeks, he treated the chair with androstenone and then analysed the lists of results that the assistants had kept for him. The chair that had once been shunned by women visitors suddenly showed a marked increase in popularity, with 21 women sitting on it. Men, by contrast, appeared to be repelled by the androstenone. Kirk-Smith's surmise had proved correct.

We now know that pheromones certainly do play a role in people's choice of partner. This finding led inventive entrepreneurs to come up with various dubious and expensive colognes containing them. Furniture manufacturers are also reputed to have used it to try to flog old-fashioned three-piece suites to women. However, its effect is only weak and in all likelihood coupled with a number of other factors.

The BBC subsequently recreated Kirk-Smith's experiment with a hidden camera for a documentary on how people choose their partners. It was at this juncture, if not before, that his experiment came to be recognized as a classic of its kind.

1976 The Razor as Teaching Aid

Bearded professors who aren't exactly on top lecturing form can pin their hopes on one thing: the electric razor. At least according to Jürgen Klapprott of the University of Nuremberg-Erlangen. For his paper 'Barba facit magistrum: a study on a bearded university lecturer's effect on his students', he taught for two semesters without a beard, then for one with a beard, and then for a final one clean-shaven again.

Professor Klapprott surmised from laboratory studies that 'fixed visual stimuli pertaining to the target person' – in other words, in this instance, beards – affected people's judgement in everyday situations about what that person was like. So even small changes in the appearance of a person's face would alter an observer's impression of it. But would this be borne out in practice?

The psychologist Jürgen Klapprott investigated the effect of his appearance on his students – with and without a beard.

In each case, ten minutes after his students had clapped eyes on him for the first time at the start of the semester, Klapprott got them to fill out questionnaires on his personality.

The results weigh heavily against university lecturers sporting beards: with a beard, the students considered Klapprott 'less purposeful, less precise, less focused, less friendly, less persistent … less able, less acute, less rational, and less intelligent'. On the positive side of the balance sheet there was little to write home about: the beard made Klapprott seem more natural, laid-back and radical – if these can be deemed positive attributes for a professor.

Source

Klapprott, J. (1976), 'Barba facit magistrum – Eine Untersuchung über die Wirkung eines bärtigen Hochschullehrers' ('The beard makes the teacher: a study on a bearded university lecturer's effect on his students'). *Schweizerische Zeitschrift für Psychologie* 35, pp. 16–27.

1976 A Millionaire Has Himself Cloned

In September 1973 an American science journalist got a mysterious phone call. When he picked up the receiver at his cabin at Flathead Lake in western Montana, the man on the other end of the line, who was reluctant to give his name, told him that he was 67 years old, rich, unmarried and in need of an heir. The caller said he was only willing to go into more detail if they could meet face to face.

This is the scene with which David Rorvik's book *In His Image* opens. It is the bizarre tale of an aging millionaire who, with the aid of researchers that the science journalist puts him in touch with, has himself cloned. However, this somewhat mediocre science-fiction story from the realm of reproductive medicine has one major flaw: Rorvik claimed that every word of it was true and that the science journalist in the book was himself.

Even before the book appeared on 31 March 1978 the press had got wind of the story. The tabloid paper the *New York Post* ran a piece on 3 March under the banner headline 'Baby Born Without A Mother: He's first human clone', telling its readers about the dawn of a new era in human reproduction. By that same evening, Rorvik's book on cloning had made it onto every television news bulletin between New York and Los Angeles.

Scientists thought Rorvik's tale of cloning was a tissue of lies. A leading mouse geneticist and embryologist Beatrice Mintz, for example, whose work was cited in his book, called Rorvik a 'fraud'. Indeed, the story of the mysterious caller, a businessman who went by the pseudonym of 'Max', was really hard to believe.

In the book Max said that he was willing to spend 'a million dollars, maybe more' to acquire a replica of himself. In

The *New York Post* of 3 March 1978: the world learned of the first cloned human being from a tabloid newspaper. But was this account really true?

NEW YORK POST

FINAL

FRIDAY, MARCH 3, 1978 25 CENTS

braces for 5-in. snow Page 2

Startling claim in book

BABY BORN WITHOUT A MOTHER

He's first human clone

By SHARON CHURCHER

A human baby, created from a single male cell in a laboratory, is now a normal 14-month-old

A Lippincott spokesman refused to tell The Post the names of any of the participants in the experiment, including the child, saying Rorvik other words, a child genetically identical to him, a twin brother shifted 70 years on in time as it were – a clone. Rorvik, who was formerly science correspondent for *Time* magazine and who was the author of several books on reproductive medicine, was meant to provide Max with contacts in the scientific community who would be prepared to attempt the experiment.

From a technical point of view, the business of cloning a human being consists of a number of extremely tricky steps. First of all, an egg needs to be removed from a woman – or preferably several eggs at once, since you can reckon on several false starts. The cell nucleus, which contains the woman's genetic material, is then taken from the ovum. The person wanting to be cloned also has to donate a cell. In principle, a cell from virtually any part of the body can be used since, with very few exceptions, each cell contains the entirety of an individual's genetic information. This genetic information is found in the nucleus, which is removed from the cell and transported into the empty ovum.

The egg assembled in this way now contains the precise genetic material from the donor. It is kept alive outside the body in a fluid culture medium until it has divided a few times and is then implanted in the womb of the woman who is to give birth to the child.

The whole procedure is fraught with problems. However, the doctor (codename 'Darwin') whom Rorvik found claimed to have solved these within 18 months, despite the fact that the world's foremost researchers in the field had failed to crack them even after decades of work. Even the technique of implanting a fertilized egg so as to ensure that it actually results in a pregnancy was only officially mastered in 1978. And that was just the first hurdle.

The greatest difficulty consisted of fusing the nucleus of a body cell with the denucleated ovule in such a way that a complete person develops from it once more. Although each cell in a person's body carries the blueprint for that individual within its genetic make-up, many genes within a cell are deactivated during the developmental process. So in a liver cell, for instance, only those genes that it requires to perform its task remain active, and the same applies to a skin cell or a brain cell.

The problem is to get the 'mute' genes to communicate again, once they have been transplanted into the empty ovum. The 'mature' nucleus has somehow to be fooled into thinking

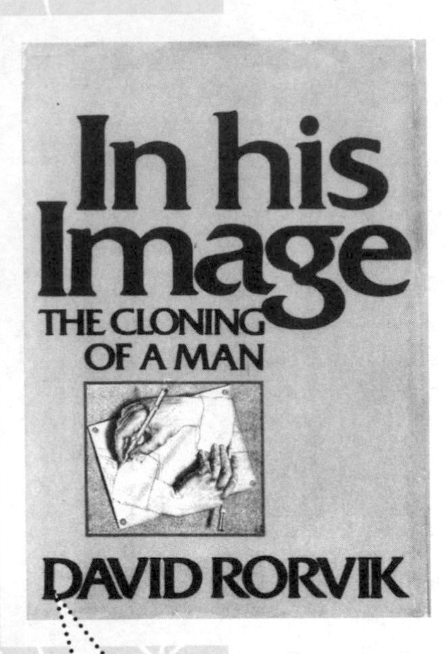

Cover of Rorvik's book. An American court found that the story was pure invention.

that it has rejuvenated, and then persuaded to set in train the development of a complete human being. Although scientists had by this stage mastered the cloning of frogs, they hadn't achieved this by using fully developed body cells, but only from immature, unspecialized cells in which no genes had yet been muted. Such cells are called stem cells.

But it wasn't just the scientific details in Rorvik's story that didn't ring true: the location and the cast didn't do much for its credibility either. Doctor Darwin was supposed to have conducted his investigations on an unnamed Pacific island somewhere beyond Hawaii, where Max owned rubber plantations and had a stake in the fishing industry. Roberto, one of Max's employees, 'with a taste for garish clothes and showy rings', set about looking 'in factories and on farms' for a suitable female candidate to carry and deliver Max's clone. Max stipulated two conditions: the woman had to be a virgin, and she had to be pretty. After an exhaustive search, a 17-year-old (codename 'Sparrow') was eventually found, who gave birth to the baby two weeks before Christmas 1976 and with whom Max promptly fell head over heels in love.

Although there's something distinctly fishy about this story, it didn't fail to cause a splash. This was a time when public opinion was taking an increasingly critical attitude towards science. Not long before, Ira Levin had published his novel *The Boys from Brazil*, about an attempt by a group of old Nazis to clone Adolf Hitler, and even some scientists had recently called for a moratorium on the newly developed technique of introducing individual genes into the genetic make-up of different organisms. Rorvik's book turned into a PR disaster for science. The German news magazine *Der Spiegel* ran a cover story entitled 'Genetics: A thousand times worse than Hitler'. For fear of providing Rorvik with even greater publicity, several scientists declined to pass any comment on the book. Others, however, were keen to stimulate a public debate. As the biologist Jonathan Beckwith of Harvard University stated: 'One day we'll wake up. Maybe it hasn't happened this time. But the next time or the time after, we'll suddenly realize we've created a monster that we didn't mean to create.'

On 31 May 1978, two months after the book appeared, a hearing took place before the US Congress on 'the area of science most properly termed "cell biology".' In actual fact, it was an investigation into Rorvik's book. Though the publisher J.B. Lippincott came under fire during the Senate hearing for having released the book, sales really took off as a result. Rorvik himself, who was summoned to give evidence, excused himself on the grounds that he was extending his promotional tour for the book in Europe.

'To protect the child from damaging publicity', Rorvik also declined to put the Senate committee in direct touch with those involved. Even the publisher didn't have a shred of proof to hand that the story had any basis in fact. Rorvik regarded the story's sheer lack of credibility as proof positive that everything was above board. An aging millionaire? A tropical island? A 17-year-old surrogate mother? – 'Put yourself in my position. Would you dare risk writing such a story? In effect, you're jeopardizing your entire career!'

And that's precisely what Rorvik did. Three months after the book was published, the geneticist J. Derek Bromhall, whose name was cited in it, filed a $7 million lawsuit for defamation of character. Max, it had been alleged, was cloned using an adaptation of the method Bromhall had developed on rabbits. In May 1977 Rorvik had sent a letter to Bromhall asking him for more information about this technique – five months after the clone had been born, according to the book. In the course of the trial, Rorvik admitted that he had invented three of the characters in his book, including Roberto. Finally, to settle the matter, he suggested a blood test, with the proviso that Max should be allowed to choose the doctors who would take samples from him and the child.

The judge declined, and ruled that a fraud had been committed. On 7 April 1982 the publisher reached an out-of-court settlement with Bromhall, which involved Lippincott paying him $100,000 and issuing a statement that it regarded the story told in the book as a fiction. As for Rorvik, he continued to maintain that his book was true.

The mystery as to why Rorvik perpetrated this deception remains unexplained to this day. People assumed that the book

Source

Rorvik, David M. (1978), *In His Image: The Cloning of Man*. J.B. Lippincott

must have a hidden political agenda or that the author was just out to make a fast buck. But there might be another explanation. A former colleague put it in a nutshell: 'David's intelligent. David's a good writer. David's also a bit strange.'

In a contribution to the online magazine *Omni* in 1997, Rorvik changed his tune, even if ever so slightly: 'I was not made privy to every aspect of the project detailed in my book, and was never provided with proof of its success ... Circumstantial evidence, nonetheless, led me to the conclusion that the project succeeded. I believed that in the late 1970s and I believe it today.' Nowadays, Rorvik fancies himself as some kind of prophet crying in the wilderness, who pointed up the possibility of human cloning long before anyone else.

Admittedly, some scientists have been a bit wide of the mark with their prognoses. For instance, when the discoverer of the genetic code and Nobel laureate James Watson was asked in an interview with *People* magazine in 1978 when he thought the first human being would be cloned, he answered, 'Certainly not in any of our lifetimes.' He went on to say, '... if either of my young sons wanted to become a scientist, I would suggest he stay away from cloning. There's no future in it.'

In 1997 the birth of Dolly the Sheep, the first cloned mammal, was announced.

On 26 December 2002, for the second time, news came through of the birth of the world's first cloned human – once again at an unknown location. According to the press release issued by Clonaid, a cloning firm founded by the Raelian UFO sect, baby 'Eve' was doing well and had the genetic make-up of the 30-something woman who donated her skin cells.

However, the gene test by an independent expert that the Raelians said would take place has been postponed indefinitely, allegedly because the baby's mother is afraid that the child will be taken away from her.

1976 The Controversy over Life on Mars

On 28 July 1976, some 205 million miles (330 million km) from the Earth, a grasping arm reached out to resolve one of the greatest questions ever posed by humanity. A small shovel on the end of the arm tipped a handful of Mars dust into a funnel leading to the biology module of the *Viking I* spacecraft. This section of the Mars lander housed three experiments, which should have provided a conclusive answer to the question of whether there is life on Mars. Instead, the results have occupied scientist Gilbert Levin's every waking (and even dreaming) moment for the past three decades. His life became one long crusade against the US space agency NASA, which is bent on suppressing the truth, or at least what Levin believes to be the truth.

As Levin explains: 'The only conclusion consistent with all the known facts is that the Viking labelled release experiment discovered microorganisms in the soil of Mars.'

Countering this, Norman Horowitz, a former colleague of Levin, who also had an experiment on board the *Viking* probe, claims, 'Every time he opens his mouth, he makes a fool of himself.' To which Levin rejoins, 'They thought Galileo was wrong the first time too.'

A technician testing the *Viking* probe's extendable shovel, which was used to collect soil samples. The results that emerged from analysing these samples are still disputed today.

NASA embarked on the project to send two probes to Mars in 1968. Eight years before, the Soviet Union had begun sending spacecraft to Mars. earlier, the long journey only yielded just so much scrap metal: virtually every craft sent to Mars disappeared without a trace into the depths of space.

This huge interest in Mars is based in particular on the fact that, of all the planets in the solar system, it is the most similar to Earth. In size, it lies somewhere between the Moon and the Earth, while its mass enables it to retain an atmosphere. It also lies at a favourable distance from the Sun. According to scientists, these are all necessary preconditions for the development of life-forms.

On top of this, thanks to the work of the eccentric American astronomer Percival Lowell a century ago, Mars had become the planet of extraterrestrials. Prompted by the discoveries of the Italian astronomer Giovanni Schiaparelli, who claimed to have seen through his telescope a network of furrows on Mars, Lowell began an intensive observation of the planet. Schiaparelli called the furrows *canali* – the Italian word meaning 'canals' as well as 'gullies' or 'furrows'. However, it was translated as 'canals', thereby evoking the idea of artificially created watercourses and hence of intelligent life-forms.

We'll never know what Lowell might have seen through his telescope without the word 'canal' at the back of his mind. But as it turned out, he ended up working out a detailed scenario for life on the planet. The Martians had built the canals, he claimed, because the planet suffered from extreme drought, thus forcing them to channel water from the polar icecaps to the equator in order to irrigate their crops.

Lowell's ideas inspired a whole generation of science-fiction writers to create highly imaginative chronicles of life on Mars. Yet, despite the fact that most astronomers refuted Lowell's thesis and couldn't make out the supposed canals when they viewed the planet through their telescopes, the existence of intelligent Martians became firmly fixed in the popular imagination.

Like all the scientists working on the Viking project, Levin didn't believe that any substantially sized animals or plants would be found on Mars. According to everything that was known, conditions were simply too harsh for such organisms to survive. If there was anything living up there, then at most it would be microbes. Yet even the discovery of such primitive life-forms would have had far-reaching implications. As one researcher put it, they would have changed the question of life on other planets 'from a miracle into a statistic'.

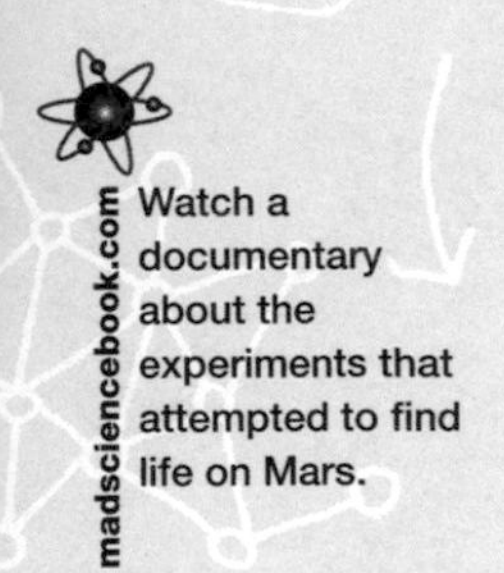

In the 1960s Gilbert Levin had devised a method that showed the presence of bacteria in drinking water or foodstuffs. NASA realized that this same procedure could be applied to the far more exciting task of searching for life on Mars.

And so it was that on 28 July 1976 a thimbleful of Martian sand was tipped into the small beaker that Levin had designed and sprayed with a nutrient solution.

Because no other form of life was known about than that on Earth, the scientists assumed that life on Mars would obey the same basic laws. The common denominator of all terrestrial life is metabolism: every life-form, from a bacterium to an elephant, metabolizes. In addition, the scientists all agreed that wherever life appeared, it would be carbon-based. As a joke, therefore, they took to calling themselves 'carbon chauvinists'.

Carbon atoms are the most versatile of all. No other atom comes close in its ability to combine to form so many diverse long-chain molecules. Proteins, hormones and the inherited material DNA – anything that is essential for life consists mainly of carbon.

Levin's calculation was simple: if the sand from the surface of Mars contained life-forms, these would take up the nutrient solution, digest it and then give off a gas. In short, they would eat. To cater for as wide a taste as possible that the microbes might have, the 'nutrient soup' was made up of a mixture of seven different molecules, most of which contained a high proportion of carbon that had been radioactively labelled beforehand (carbon-14). A Geiger counter would then measure whether this radioactive carbon reappeared in the gas that came out of the container. This would have been a clear indication that something in the sand had consumed the nutrients and released the decomposition products in the form of gas.

Source

Ezell, E.C., and Ezell, L.N. (1984), *On Mars. Exploration of the Red Planet 1958–1978*. The NASA History Series (SP-4212),

Alongside Levin's experiment in the biology module on board the *Viking* probe were two other experiments investigating other aspects of metabolism. It was as if three well-appointed university laboratories had been compressed to the size of a car battery. A remotely controlled clockwork mechanism consisting of 4000 parts was intended to answer the greatest question of all. This component alone of the *Viking* probe cost $59 million, while the entire programme cost close to $1 billion.

When Levin received the first results after two days of waiting, he couldn't believe his luck: the Geiger counter was ticking like mad! The gas was clearly radioactive, providing apparently conclusive proof of life on Mars. Whatever there was in the Martian sand, it was far more active than fertile soils on Earth teeming with bacteria. Levin signed the first page of the data sheets, convinced that this piece of paper would go down as a key document in the annals of mankind.

The second of the three experiments, which tested whether there was anything breathing in the sand, also yielded a positive result. On 31 July the biology team announced at a press conference that the reaction observed in Levin's experiment appeared to be 'very much like a biological signal', but at the same time warned against drawing hasty conclusions, since the data were almost too good to be true: the soil reacted simply too quickly. Microbes normally take some time to take up the nutrient solution, digest it and give off the products of decomposition, and yet the two experiments conducted thus far showed an instant reaction.

And so, precisely what the geologists on the Viking project had long since predicted came to pass, namely, that the biologists would have no way of evaluating the results of their experiments. The geologists, who had devised their own set of instruments for the *Viking* probe, were the natural enemies of the biologists. Even before the project began, both teams were already vying for the lion's share of the maximum permitted payload and, once the probe had landed, for as much radio time as they could get to transmit their data back to Earth.

Although the geologists had constantly warned that not enough was known about Mars to be able to answer the question about life existing there from a single experiment, the biology experiments were the driving force behind the *Viking* programme right from the outset, as far as political and public interest in the project was concerned. The geologists were very much the second-class passengers on board *Viking*.

Then a third experiment – one that wasn't part of the biology module – really put the cat among the pigeons when it failed to detect any organic compounds in the Martian soil. Organic compounds are those long-chain carbon molecules that scientists regard as the prerequisite for life. They had actually expected *Viking* to find organic compounds in the Martian soil but no evidence of life. Now they were faced with the exact opposite: Levin's experiment indicated that life was present, but there was no sign of organic compounds.

On 3 September 1976 the sister probe *Viking 2*, identical in its construction, landed on Mars. It was able to repeat the experiments, but this didn't resolve the matter. Rather, the scientists fell to arguing about where the grab should take the soil from, who was entitled to conduct which experiment, and how the data were to be evaluated.

Before long, the suspicion arose that, in searching for an exotic biology, what the *Viking* scientists had actually stumbled upon was an exotic chemistry that had nothing to do with life. Nowadays, the assumption is that substances such as hydrogen peroxide fooled the boffins. If they come into contact with water and metals, these can release gases that could be interpreted as signs of life.

Subsequently, a whole battery of tests was undertaken in laboratories on Earth to try to reproduce the *Viking* results. Today, most scientists are convinced that chemical reactions really were what lay behind the supposed signs of life from Mars.

As for Levin, he still holds fast to his interpretation of things. He maintains that, in looking at pictures of rocks on Mars, he has seen green patches that move. In his campaign against NASA he is supported by conspiracy theorists, who have known for a long time that the government is hiding something from them.

To settle the matter once and for all, Levin has proposed countless new experiments to NASA. But the only reaction there whenever his name is mentioned is a deep sigh.

1977 Country and Western Philosophy

It doesn't often happen that a country-and-western song acts as a spur to science. Yet Mickey Gilley's 'Don't the Girls All Get Prettier At Closing Time?' has spawned quite a few scholarly articles. The person who's to thank for this is the psychologist James W. Pennebaker; the work that he wrote when he was at the University of Virginia – entitled 'Don't the Girls Get Prettier at Closing Time: A Country and Western Application to Psychology' – is in the running for being the funniest ever description of an experiment. In this paper, Pennebaker wrote that 'The jukebox ... has long been a rich source of social psychological truths. Many a hypothesis is available to the researcher for a mere quarter (three hypotheses for 50 cents).' It was a common prejudice, he continued, that country music was only about 'mama or trains or trucks or prison or getting drunk'. On the contrary, it was positively chock-a-block with psychological themes. Take, for instance, the sense of justice evident in Hank Williams's 'Your Cheatin' Heart Will Tell On You', the dissonance theory in Johnny Cash's 'A Boy Named Sue' or Skinner's notion of positive reinforcement (see p. 97) in Lefty Frizzell's 'If You've Got the Money, Honey, I've Got the Time'.

When Pennebaker heard 'Don't the Girls All Get Prettier At Closing Time?' for the first time, he realized straight away that it not only embodied 'reactance theory and predecisional judgements of attractiveness of alternatives', but that the song also suggests a method by which the hypothesis contained in the title can be tested.

Source

Pennebaker, J.W. (1979), 'Don't the Girls Get Prettier at Closing Time: A Country and Western Application to Psychology'. *Personality and Social Psychology Bulletin* 5, pp. 122–125.

In October 1977 six of Pennebaker's students went round Charlottesville's three bars with questionnaires and asked the patrons at 9.00 p.m., 10.30 p.m. and midnight how attractive they found members of the opposite sex in the bar, on a scale of 1 to 10. The bars closed at half-past midnight.

The findings showed that the Mickey Gilley's song was right. The later it got, the prettier the men in the bar found the women. And vice versa. One possible explanation for this is provided by dissonance theory: any man who doesn't want to go home alone would be stupid to find all potential partners unattractive. This dissonance is removed when his critical judgement of attractiveness becomes more tolerant as the hour of truth approaches. Gilley was wise to this as well: 'Ain't it funny, ain't it strange, the way a man's opinions change, when he starts to face that lonely night?'

This experiment has frequently been repeated, with mixed results. One study differentiated between the results and answered Pennebaker in the journal *Basic and Applied Social Psychology* with an article entitled 'They Do Get More Attractive at Closing Time, But Only When You Are Not in a Relationship'. This same study also explored the phenomenon of why some people are shocked when they wake up the next morning: 'one's relative satisfaction (or regret) with one's choice when seen in the "light of day" may be based on a different assessment of attractiveness ...'.

Pennebaker published his study in the *Personality and Social Psychology Bulletin*. The renowned *Journal of Personality and Social Psychology* turned it down for publication, justifying their decision in terms that Pennebaker remembers fondly as 'one of the neatest rejections I ever got'. The journal's editor simply wrote: 'We Ain't Going to Take It (The Who, 1968)'.

1978 Want to Come to Bed with Me?

The 16 women who were staying on the Florida State University campus in Tallahassee found themselves on the receiving end of a very direct chat-up line. A young man approached them and, without more ado, said, 'I have been noticing you around campus. I find you to be very attractive. Would you go to bed with me tonight?'

All the women declined, responding either by saying 'You've got to be kidding', or 'What is wrong with you? Leave me alone.'

Out of the 16 men who were approached in the same way by a woman, 12 took up the offer. Their responses were 'Why do we have to wait until tonight?' or 'I can't tonight, but tomorrow would be fine.'

The psychologist Russell Clark, who carried out this experiment, wanted to find out how the sexes responded differently to sexual advances. His findings were clear. However, it was 11 years before they were published.

The 1970s were a period of social upheaval. The idea that men and women differ from one another right from birth not only in their physique but also in their behaviour was decried as a chauvinistic attitude that existed solely for the purpose of denying women equal rights. Anyone who claimed that men and women approached their choice of partners differently on biological grounds (a fact that Clark was sure was true) was viewed with suspicion by many social psychologists.

The idea for the experiment came from his seminar in social psychology, in which Clark discussed a paper by James W. Pennebaker that had recently appeared: 'Don't the Girls Get Prettier at Closing Time: A Country and Western Application to Psychology' (see p. 253).

With regard to this study, Clark lighted on the topic of the differences between men and women in choosing a sexual partner: 'A woman, good looking or not, doesn't have to worry about timing in searching for a man. Arrive at any time. All she has to do is point an inviting finger at any man, whisper "Come on 'a my place," and she's made a conquest. Most women can get any man to do anything they want. Men have it harder. They have to worry about strategy, timing, and "tricks".' The women in the seminar protested at this. To which Clark simply said, 'We don't have to fight. We don't have to upset one another. It's an empirical question. Let's design a field experiment to see who's right!'

A few weeks later, five women and four men were roaming the university campus, trying to make contact with the other sex. Alongside the blatant offer of sex, they also tried two alternative strategies: 'Would you go out with me tonight?' and 'Would you come over to my apartment tonight?' The same number of men and women took up the first offer, namely half of those asked. But only 1 woman out of 16 agreed to come to the apartment, while

11 of the 16 men were game for it. All the women turned down the offer of sex, while 12 men jumped at it – in other words, half as many again as were prepared to just meet up with a woman on normal terms.

Clark was certain that the reason for this difference lay in the asymmetrical biology of the sexes: 'In order to produce a child, men need only to invest a trivial amount of energy, a single man can conceivably father an almost unlimited number of children. Conversely, a woman can give birth to and raise only a limited number of children.'

The different cost of sex for men and women was, then, a direct cause of the behaviour that Clark had noticed in his experiment. Women are selective, whereas men are basically prepared to go to bed with any woman. In contrast to the women, who all reacted with outrage to the offer of sex, the four men who didn't accept were concerned to give the women an excuse, either by saying 'I'm married' or 'I'm going with someone'.

When Clark tried to publish his study, he was left in no doubt that his findings didn't exactly chime in with the mood of the age. From one journal, he even got the following dusty response: 'This paper should be rejected without possibility of being submitted to any scholarly journal. If *Cosmopolitan* won't print it, [...] then *Penthouse Forum* might like it. But not —— [name of journal omitted].'

Later, the psychologist Elaine Hatfield got to hear of the experiment and tweaked Clark's article slightly before resubmitting it herself. Journal editors' reactions were somewhat more muted and their rejections more convoluted: 'I feel the paper should (and almost certainly will be) published somewhere. I regret that I cannot tell you we will publish it.'

Source

Clark III, R.D., and Hatfield, E. (1989), 'Gender Differences in Receptivity to Sexual Offers'. *Journal of Psychology & Human Sexuality* 2 (1), pp. 39–55.

Then they found a new point of criticism: the findings were out of date. Perhaps the difference between the sexes had been like this in 1978, but things had changed in the interim. So Clark repeated the experiment in 1982 – with practically the same result. After further rejections, the study was finally published in 1989 by the *Journal of Psychology & Human Sexuality*. When the suspicion arose that the fear of AIDS might have changed sexual behaviour, Clark sent his students out into the field once more – with the familiar result.

Nowadays the study 'Gender Differences in Receptivity to Sexual Offers' crops up with great regularity in the media under various headlines ('Indirect evidence that men are stupid', 'Guys = Icky: The definite proof'). The BBC repeated the experiment for a documentary film involving a hidden camera. It turned out that Englishmen were icky as well.

1979 Free 'Unwill'

A second is a long time. Too long, as far as Benjamin Libet was concerned. The American neurological researcher heard about this particular second for the first time at a scientific conference in May 1977, 12 years after it had been measured. A second is the time that elapses, during a voluntary hand movement, from the first moment the brain plans the action to the execution of the actual movement, according to a paper that Hans Kornhuber and Lüder Deecke published in 1965. At the time, these German neurologists had discovered electrical changes that take place in the brain just before an action, a phenomenon that they christened 'readiness-potential' (*Bereitschaftspotential*).

It was no surprise to learn that readiness-potential kicks in just before any movement – after all, muscles can only become active once they have received instructions from the brain to become so. Even so, in one particular regard the result was absurd.

The test subjects were left to decide for themselves when they should move their hand. This meant that there had to be at least one second between the moment they made this voluntary decision and the instant they moved their hand. It struck Libet straight away that this contradicted people's everyday experience: one second between the decision to pick up a pencil and actually picking it up was clearly too long.

These calculations had been based on a premise that seemed so self-evident that nobody had ever taken the trouble to check it, namely, that the conscious decision to make a movement must occur before the brain makes the first moves to perform it. Simple cause and effect. Nobody could seriously doubt that this was the case – or could they?

Libet wanted to measure the time lapse precisely. 'The whole of the following year I asked myself how on earth I might measure the instant of conscious decision.' Kornhuber and Deecke had captured only the moment of readiness-potential and that of the movement, but not the instant of conscious decision, since

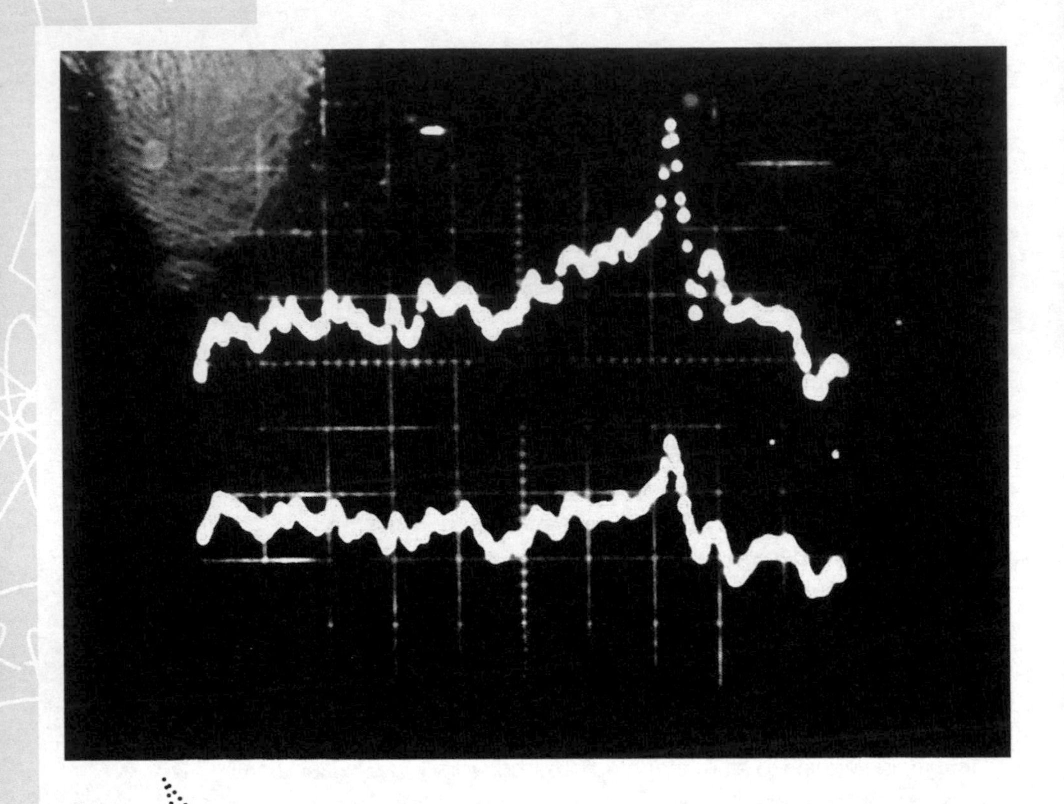

These measurements of brain impulses, which precede a movement, led scientists to speculate that there may be no such thing as free will.

the only person who is party to this is the test subject herself. Because it couldn't be objectively measured or identified from electrical brain impulses, researchers had given it a wide berth. Free will was regarded as not being susceptible to scientific investigation. As Libet said, 'I do believe people are genuinely afraid of it.'

And so Libet looked for a possible way in which the test subjects might be able to communicate to him when they decided to move their hand. However, they could neither say anything nor make a gesture, since these signals would themselves have been affected by that same unconscious delay that attends every willed action.

Then Libet suddenly got the idea of using a clock. If the test subjects glanced at a fast-moving clock and noticed when they took the decision to make a movement, then they could subsequently let the experimenter know when that moment was. At first, Libet had doubts about his brainwave: 'Because the measurement had to be very exact, I didn't think it would work, but I decided to try it anyway.'

No study in the history of the neurosciences has stirred up so much controversy or so many differing interpretations as these tests, for what Libet discovered was that there may be no such thing as free will.

In March 1979 the first of five test subjects, the female psychology student C.M., took a seat in the comfortable armchair in Libet's laboratory at the Mount Zion Hospital in San Francisco. Electrodes were attached to her head and her right wrist, and she was asked to look at a small screen about 6 ft (1.8 m) in front of her. On it, a green dot was revolving in a circle, taking 2.56 seconds for each complete revolution; this was to serve as the clock. Libet now asked C.M. to bend her right wrist at a moment of her choosing. The change in voltage through the electrode on the subject's wrist told him the precise instant at which the movement had occurred, the readiness-potential was given by the electrodes on the subject's head, and the moment of conscious decision was revealed to him after each phase of the test by C.M. herself, who noted where the green dot had been when her free will kicked in.

'The test subjects had no idea what was going on and found it all rather curious,' Libet recalls. But for $25 per session, they were quite content to move their wrist at a moment of their choosing.

'Even after the first test, I noticed how unusual the result was,' says Libet, who is now aged 85, pulling his old laboratory notes out of a drawer. These consist of a stack of papers with figures scribbled untidily all over them, interleaved with photos of graphs on a screen: the readiness-potentials.

The moment that C.M. gave as the instant in which she made her decision to move invariably fell just 0.2 seconds prior to the actual movement. This was a rational result that accorded with everyday experience. But the readiness-potential occurred at least 0.55 seconds – and in some instances, as in Kornhuber and Deecke's study, even a whole second – before the movement. In other words, C.M.'s brain set in train an action that the brain actually still knew nothing about, since C.M. would only decide to take that action a third of a second later. It was just the same with the other test subjects: the readiness-potential was always present long before the onset of free will.

Source

Libet, B., et al. (1983), 'Time of Conscious Intention to Act in Relation to Onset of Cerebral Activity (Readiness-Potential). The Unconscious Initiation of a Freely Voluntary Act'. *Brain* 106 (Pt 3), pp. 623–642.

At first sight, the experiment seemed to point to only one conclusion: that free will is an illusion. The brain puts up consciousness as a stalking horse to fool us into believing that we have a free choice. But in the depth of our subconscious, things are already a done deal. We don't do what we want, rather we want what we do.

Libet wasn't happy with this conclusion, however. 'This meant that we were essentially nothing but sophisticated automata, with our consciousness and our intentions just some tacked-on by-product without any causal force.' In suggesting this, the experiment undermined the very foundations of our legal system. Could any court punish a person for an act that he actually couldn't have helped but commit?

Immediately, therefore, Libet came up with a new theory: although the experiment did indeed demonstrate that we had no power over those designs that arose from the subconscious masquerading as free will, we did have the power to intervene. In a series of further experiments, Libet proved that the two-tenths of a second between the conscious decision and the action was a sufficient lapse for us to put a veto in place and abort the whole show. If we didn't exactly have a free will, then we at least had a free 'unwill'.

This finding also chimed in with religious and ethical standards that exhort us to exercise self-control, as well as with the Ten Commandments, many of which begin with the words 'Thou shalt not ...'. As Libet was fond of quipping, his veto theory even provided a 'physiological explanation for original sin': 'If a person already regards the evil intention as sinful, even if it doesn't result in any action, then this effectively makes all men sinners.'

However, the veto theory has one crucial weakness: if a conscious decision is preceded by an unconscious brain impulse, then why shouldn't this also apply to Libet's conscious veto?

Some scientists believe that Libet was concerned to salvage the idea of free will because he was afraid of the logical conclusion of his own experiment. For instance, the philosopher Thomas W. Clark wrote: 'His underlying reasoning is this: because it is unthinkable that we do not possess free will (after all, we don't want to be automata, do we?), then we should waste no time in discovering a proof for the existence of free will.' Such reasoning, though, is unscientific.

This argument also begs the question whether there is a non-material spirit, or is consciousness solely the result of chemical and physical processes in the brain? In the second eventuality – a position taken by determinists – Libet's experiment isn't so remarkable. If the spirit consists of a series of material reactions taking place in the brain, then free will must have been initiated by unconscious brain impulses. Otherwise it is just not possible. Every effect must have a cause.

Viewed in this light, then, Libet's findings aren't so spooky after all. They simply run counter to our personal experience. We like to think that we have free will, and so we believe this to be the case. Even neuroscientists can't help feeling this. While many of them claim to have put aside all notions of personal guilt and atonement, they have to admit that they cannot resolve the contradiction between scientific knowledge and personal feeling in their daily lives.

So, despite the fact that he doesn't believe in free will, the German neuroscientist Wolf Singer still acknowledges that 'I go home in the evening and blame my children for having done something stupid, because I naturally assume that they could have acted differently.'

1984 Bigger Tips for Touching

Research into people's tipping habits may not be able to compete in terms of kudos with other areas of science, but its findings do have the rare distinction of being relevant to people's daily lives. Another reason for its popularity is that it enables behavioural studies to be carried out simply and cost-effectively: there are any number of restaurants, the customers serve as test subjects, and if you want to find out the effect of experimental manipulation, all you need to do is count how much you've received by way of tips.

April H. Crusco, a student at the University of Mississippi, wasn't actually concerned with the science of tipping; rather, her field of study was the significance of touch in a therapeutic context. However, a field study in this realm would have been too complicated, and because she also worked on the side as a waitress, she hit on the idea of employing her fellow waitresses in the restaurant as surrogate therapists. If touching was effective as part of a therapy – say, as a way of invoking another person's sympathy or asserting your own power – then it should by rights also have an effect in a restaurant. And this effect should be apparent in the size of the tip.

The first thing that Crusco and Christopher G. Wetzel, whose social psychology class she was attending, worked out was how the waitresses should touch the customers. They plumped for two variations that were easy to execute and which at the same time appeared entirely natural. One of these experimental conditions was dubbed 'fleeting touch', and involved the waitress touching the customer's hand twice with her fingers for half a second at a time. Crusco expected that this would have a positive effect. The second experimental condition, which they called 'shoulder touch', entailed the waitress resting her hand on the customer's shoulder for one and a half seconds. This form of touching could be interpreted as dominant behaviour, and Crusco predicted that it would have a negative effect.

The waitresses practised both forms of touching until they could do them really casually, without arousing any suspicions. Anyone wishing to repeat the experiment will find the precise sequence of actions detailed in the paper that came out of it: 'The waitress approached the customers from their sides or from slightly behind them, made contact but did not smile as they spoke "Here is your change" in a friendly but firm tone, bent their bodies at approximately a 10 degree angle as they returned the change, and did not make eye contact during the touch manipulation.'

Such treatment certainly left its mark on 116 customers at two restaurants in Oxford, Mississippi. The fleeting touch on the hand yielded an average of 37 per cent more in tips, while the shoulder touch – counter to expectations – still managed to increase tips by 18 per cent.

For all those waitresses who have an aversion to touching their customers, here are a few more ploys that also had a positive effect on the level of tipping in later experiments: introducing yourself by your Christian name; squatting down while taking people's orders; writing a handwritten 'Thanks!' on the bill, or drawing a sun or smiley face on it; repeating the customer's order; and enclosing a joke with the bill. The particular joke used in the study was this: 'An Eskimo had been waiting for his girlfriend in front of a movie theatre for a long time, and it was getting colder and colder. After a while, shivering with cold and rather infuriated, he opened his coat and drew out

Source

Crusco, A.H., and Wetzel, C.G. (1984), 'The Midas Touch: The Effects of Interpersonal Touch on Restaurant Tipping'. *Personality & Social Psychology Bulletin* 10 (4), pp. 512–517.

a thermometer. He then said loudly, "If she is not here at minus 10, I'm going!"' In return for such a lame joke, customers were prepared to leave 50 per cent more tip.

1984 Efficient Chat-up Lines

When professor of psychology Michael R. Cunningham lighted on the subject of the attraction between the sexes in a social psychology seminar, his students asked him what, from a scientific point of view, the best chat-up lines were. Cunningham sifted through a stack of learned journals and finally came across a paper that listed 100 chat-up lines in order of their popularity and that organized them into three categories: direct, innocuous, and cute/flippant. However, this hit parade was merely the upshot of a form-filling survey, meaning that the 100 opening gambits had never been used in earnest. Cunningham resolved to change that.

So, a couple of weeks later, an averagely attractive man tried it on with unaccompanied women in a bar in Chicago, using six lines from the three categories. The direct approach was represented by 'I feel a little embarrassed about this, but I'd like to meet you' and 'It took a lot of nerve to approach you, so can I at least ask what your name is?' The innocuous approaches were either simply 'Hi!' or 'What do you think of the band?' Finally, the cute/flippant approaches were either 'You remind me of someone I used to date' or 'Bet I can outdrink you.'

Cunningham sat and watched from a short distance away and noted the effect. A smile, eye contact or a friendly response meant that the chat-up had been successful. Turning away, walking out, or a dismissive reply signalled a failure.

The strike rate was greatest with the direct method: 9 out of 11 women reacted favourably to the 'I feel a little embarrassed ...' spiel, and 5 out of 10 to 'It took a lot of nerve ...'. The innocuous approaches were similarly successful. By contrast, it seems that cute/flippant chat-up lines are to be avoided at all costs: 80 per cent of women responded negatively to them.

In further tests, Cunningham found that women thought that the cute/flippant remarks pointed to negative personality traits in the man, such as shallowness or an overbearing nature.

Source

Cunningham, M.R. (1989), 'Reactions to Heterosexual Opening Gambits: Female Selectivity and Male Responsiveness'. *Personality and Social Psychology Bulletin* 15, pp. 27–41.

When Cunningham repeated the same experiment in reverse, he had a fair idea what the result would be even before he started. Men reacted in similar fashion to every type of chat-up line that women used, namely 80–100 per cent positively.

1984 The Voluntary Stomach Ulcer

Barry Marshall didn't even bother to seek approval for his experiment; he knew he'd never get it. Nor did he say anything to his wife about the strange concoction that he swallowed on that Tuesday morning at 11 o'clock. It was 10 July 1984, and in his laboratory at Fremantle Hospital in Western Australia, Marshall had just mixed around a billion bacteria from the stomach of a 66-year-old man with a little water. 'It smelt a bit disgusting, rather like fresh meat,' recalls Marshall. So little was known about the bacteria in his cocktail that they didn't even have a name. All that the 33-year-old Marshall hoped was that they would make him really unwell.

Three years before, as a trainee doctor, Marshall had been searching around for a research project – a task he was required to complete as part of his medical training. At the Royal Perth Hospital, he got to know the pathologist Robin Warren, who had discovered an unknown bacterium in tissue samples from patients with gastritis. Marshall tested further samples and saw that most were infected. In the library he read with astonishment that Warren wasn't in fact the first to have noticed these bacteria. As early as the 19th century, researchers had found spiral-shaped bacilli in the stomachs of humans and animals. Might these bacteria have something to do with the gastritis?

Marshall treated his first patient with antibiotics, which successfully got rid of the unknown germs as well as the gastritis. The results of this test confirmed Marshall's belief that the bacteria caused not only gastritis but also duodenal and stomach ulcers.

Marshall's colleagues advised him to keep this suspicion to himself. For one thing, he hadn't even finished his training, for another he didn't have any proof for his thesis, and thirdly his bacterium hypothesis flew in the face of accepted wisdom. Prior to this, stomach problems had been attributed to psychological problems and stress. No other physical ailment was so strongly associated with frustration, anxiety and emotional instability as the stomach ulcer.

But Marshall was far too fired up to keep a lid on his ideas. 'I had nothing to lose, I wasn't some eminent academic who was having to defend 20 years of research.' In September 1983 he presented his findings at the Second International Workshop on Campylobacter Infections in Brussels. With his missionary zeal and his self-confidence, which bordered on arrogance, he immediately attained a kind of controversial celebrity. Many of the audience found that his presentation was singularly lacking the appropriate degree of modesty and restraint.

The very idea that bacteria could trigger gastritis was a bold assertion in itself, but to claim that bacteria could live for months, even years, in the stomach was completely daft. The two litres of gastric juices that a human being produces every day consist in large measure of hydrochloric acid and can dissolve a nail. A thick layer of mucus therefore exists to protect the stomach from digesting itself. It was impossible that germs could survive in such an environment.

Physician Barry Marshall infected himself with bacteria to prove that they were the cause of stomach ulcers.

The experts assumed that the early studies that Marshall had found had reached the wrong conclusions as a result of using contaminated probes. And even if such bacteria were present in the stomach, that didn't remotely prove that they were the cause of illness. It seemed more likely that the bacteria colonized the lesions in the stomach only after they had formed of their own accord.

Marshall knew which element of proof he still needed, namely, the last two of the four postulates that the German physician Robert Koch had established for identifying a germ as a pathogen:

1. The bacterium must occur in every case of the disease.
2. The bacterium must be able to be cultivated outside the body.

3. The disease must be able to be induced in experimental hosts by these cultivated bacteria.
4. The bacterium must in turn be able to be extracted from the experimental hosts and cultivated once more.

The first postulate posed no problems. Time and again, Warren and Marshall had found the bacteria on their patient's stomach lining. Postulate number two was somewhat more difficult to fulfil. Marshall's colleagues tried for months on end to cultivate the bacteria in the clinic's laboratory, but to no avail. Normally, bacteria take no longer than two days to multiply in a Petri dish. If you let them grow for any longer, they completely overrun the culture medium they're being grown on. But in the case of Marshall's bacteria, there wasn't the slightest sign that they were multiplying, even after 48 hours.

Around 30 fruitless tests were conducted before a dangerous infection that broke out among both patients and staff at the clinic in Easter 1982 suddenly provided an unexpected breakthrough. Because of a shortage of staff, Marshall's project was put on a back burner during this period, and as a result his culture dishes remained in the warming cupboard longer than the customary two days. After five days, as if by magic, the bacteria had multiplied.

However, the real difficulties began with the third postulate – the introduction of the disease into a healthy organism. Two rats into whose stomachs Marshall injected the bacteria developed no symptoms, while two piglets 'fought off the infection'.

If the infection couldn't be proved using a test animal, then the only route left to Marshall was epidemiological studies in humans. This involved deducing the cause of the illness through statistical analysis from the data of as many patients as possible with stomach problems. But this could take years before a clear picture emerged. Marshall didn't want to wait that long, since he knew that obtaining the proof with a suitable experimental host would only take a matter of weeks. There was only one thing for it: he'd have to play the role of guinea pig himself.

For the first few hours after drinking the bacterial soup, he noticed an 'increased abdominal peristalsis (audible gurgling at night)'. Thereafter, nothing more happened for a week. On the morning of the eighth day, Marshall vomited a small amount of mucus. In the second week of the experiment, Marshall's mother noticed that he had halitosis. He was also experiencing headaches and was irritable. On the tenth day, a colleague finally introduced the flexible tube of the gastroscope down Marshall's gullet into his stomach in order to take two samples. Marshall had already endured this same procedure five weeks previously. He had done so at that time to ascertain that his stomach was perfectly healthy prior to experimenting on himself.

This new sample was dyed and placed under the microscope. The cells in the upper layer of skin were damaged, and white blood corpuscles had gathered on the mucus. Marshall had gastritis – Koch's third postulate was fulfilled.

For the fourth postulate, Marshall isolated the bacteria from the second sample and cultivated them on a culture medium. They were the same as those that he'd swallowed ten days earlier. Now there was no longer any doubt: the bacteria (which later came to be called Helicobacter pylori) were capable of causing gastritis. Marshall was over the moon. 'I hoped that an ulcer would develop from the gastritis, which would provide me with material for papers for years to come.' But when he told his wife about the experiment, she gave him the stark choice of either taking antibiotics or moving out and finding his own flat. Marshall opted for the antibiotics; however, these weren't actually necessary. The infection wore off of its own accord after two weeks. Evidently Marshall's immune system had seen off the interlopers. This was consistent with the way in which this bacterium spread. Nowadays, it's reckoned that around half the world's population is infected with it, but that only a tiny fraction of them come down with gastritis or stomach ulcers.

Heliobacter pylori bacteria can survive in the hostile environment of the stomach.

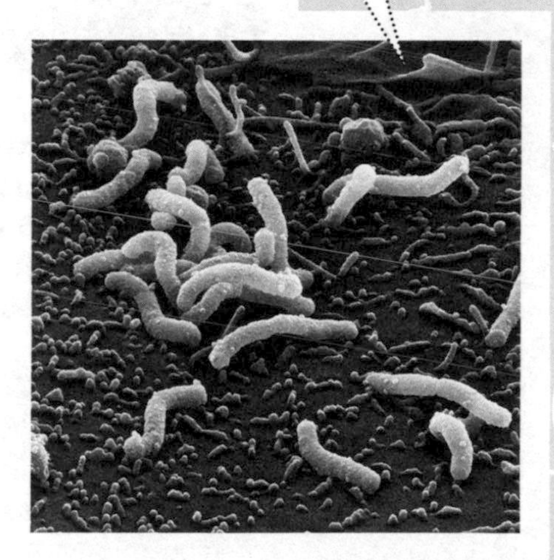

The public got to learn of Marshall's experiment not in the usual way, via a medical journal, but directly from the American tabloid newspaper the *Star*. Shortly after the matter was done and dusted, Marshall got a call from a reporter who wanted to interview him about an earlier article he'd written. Marshall couldn't keep his mouth shut and so ended up, as the 'Guinea-Pig Doctor', sharing the pages of the *Star* with Lady Di and the latest celebrity diets.

However, another ten years were to pass before the news that stomach ulcers are infectious diseases filtered down to

Source

Marshall, B.J., et al. (1985), 'Attempt to Fulfil Koch's Postulates for Pyloric Campylobacter'. *The Medical Journal of Australia* 142 (8), pp. 436–439.

the general run of doctors. On the one hand, the pharmaceutical industry had very little interest in spreading the news that antibiotics can get rid of gastritis for good within weeks. They were doing very nicely indeed out of selling antacids that in some cases had to be taken for years. On the other hand, Marshall had only fulfilled Koch's four postulates for gastritis, not for stomach ulcers. Although public health authorities now recommend that one should treat stomach ulcers with antibiotics, a number of eminent experts still find themselves at odds with Marshall's ideas.

Marshall, who was awarded the Nobel Prize for Medicine in 2005, set a trend with his experiment for reassessing other diseases as infectious. Today, researchers are busy speculating about the possible influence of bacteria and viruses on schizophrenia, heart attacks, rheumatism and diabetes. However, few of these conjectures have been proved thus far.

1986 A Year in Bed

It sounds like the ideal job for sluggards: the 11 men who were selected to take part in this experiment in January 1986 were required to go to bed and lie there – for a whole year. They spent 370 days and nights in this position, without ever getting up or even sitting up. They were washed lying down, and ate, read, watched television and wrote letters in a prone position. Boris Morukov from the Institute for Biomedical Problems in Moscow wanted to find out what happens to a person during a long journey in weightless conditions. Morukov is a physician and a cosmonaut.

Bed-rest studies were instituted in the 1960s, as astronauts began to spend ever longer periods in space. Before long, the question arose as to what effect the lack of gravity has on the human body. Since there is no possibility on Earth of keeping a body in weightless conditions for a protracted period (see p. 132), the effect had to be simulated. And the simplest method of doing so was to place a test subject in a bed that was tilted back towards the headboard at an angle of 6 degrees.

Such a position has a similar effect on the body to weightlessness: the heart no longer has to work against gravity and so switches to a slower rate, the muscles and skeleton have hardly any load on them and are partially degraded, and the red blood cell count decreases, because the body is doing less work and so requires less oxygen. The first bed-rest studies lasted a few days, but later ones extended to a few weeks or even two or three months. But the 370 days of the Moscow study went way beyond anything that had been done before.

It's hard to say what induced the 11 men to take part in the experiment. Was it, as Morukov believes, the urge to make a contribution to science? Or the decorations that the Soviet state dished out for such achievements? Or maybe the car that each of them had been promised? As Morukov says, 'It was still the Soviet era then, and getting hold of a car wasn't easy.' In any event, the participants took the study very seriously. Only one of them quit the experiment, after three months – he already owned a car.

The physician and cosmonaut Boris Morukov conducted the longest bed-rest study in the history of medicine.

The purpose of the experiment was to test new ways of preventing the body from degenerating. The test subjects did weight training exercises while lying down or went for a walk on a vertical treadmill placed in front of the bed. Five of the men were only allowed to do these exercises after four months in bed. This was designed to simulate the eventuality of training being suspended for a long period in space due to illness or a loss of power in the spacecraft.

After four months, eight months, and at the end of the study, the men were put into a centrifuge as they lay on their beds and subjected to eight times the G-force they would normally experience on Earth. This is the kind of acceleration that is encountered at the end of a spaceflight when the capsule re-enters the Earth's atmosphere. Once the year was over, there was a two-month period of rehabilitation, during which the bed cosmonauts had to learn again how to sit and walk.

Greater than the strain on their bodies was the psychological stress. The men were put in groups in three different rooms, where they spent their time

Source

Grigorev, A.I., and Morukov, B.V. (1989), '370-Day Anti-Orthostatic Hypokinesia'. *Kosmicheskaia biologiia i aviakosmicheskaia meditsina* 23 (5), pp. 47–50.

watching TV and reading. Initially, they planned to learn a foreign language during their stay, but gave this up after two weeks. Even having their food served, in proper spaceman fashion, in aluminium canisters didn't brighten their mood. But at least it helped furnish them with a hobby: shackled to their beds, they started making ships from the aluminium containers, or medals for the nurses. They made a knight in shining armour as a present for Morukov. On their birthdays they gave each other presents, while they celebrated public holidays by throwing parties, in so far as they were able to while lying flat.

The boredom and the constant medical examinations also led to tensions. The occupants of one five-man room fell out with one another so badly that one of them had to be moved out. 'Otherwise, something unfortunate might have happened,' recalls Morukov, who also replaced medical personnel whom the men didn't get on with. 'The one thing I couldn't do without was the men.'

The participants were aged between 27 and 42, and several of them were doctors themselves. Most had wives and children, whom they only got to see once a week, on Sundays. Some marriages didn't survive the strain. And one of the men fell in love with a researcher who was working on the project.

1992 Doing the Deed in an MRI Scanner

It was the most unusual place that Ida Sabelis had ever had sex. On 24 October 1992 this 40-year-old Dutch woman and her partner Jupp found themselves lying – both naked – on the examination bed of the Groningen University Clinic's magnetic resonance imager. A radiologist slid the bed into the tunnel of the machine – a space barely 20 inches (50 cm) wide – and left the room. An improvised curtain covered the window to the control room, where the physicians Willibrord Wijmar Schultz and Pek van Andel sat waiting for a historical image to materialize on their screen. They had set the machine for a combined patient weight of 330 pounds (150 kg).

As early as the end of the 15th century, Leonardo da Vinci had made a sketch of a couple having sexual intercourse. This cutaway drawing gave a glimpse into the human body and showed the anatomy of the penis, vagina, womb and other internal organs. Da Vinci doubtless based his sketch on corpses he'd seen being dissected. However, because the dead don't have sex, he'd had to guess at what happens to certain organs during copulation.

The next serious attempt to make a drawing of the sex act was published in 1933. The sex researcher Robert Latou Dickinson gleaned his knowledge from experiments in which a test tube around the size of an erect penis was inserted into sexually aroused women. Later, researchers like Alfred C. Kinsey, William Masters and Virginia Johnson experimented with artificial penises and speculums. Yet ultimately the insights they gained from these didn't tell them much about what a person's insides really look like during intercourse.

In 1991 the physician Pek van Andel happened to see an MRI scan of a singer's epiglottis taken while he was humming a note. The image reminded him of Leonardo da Vinci's drawing and he began to wonder whether he might be able to capture such an image during coitus. To do this, he first needed to get hold of a magnetic resonance imager. However, the hospitals he approached didn't take his request seriously. But what he did find out was that MRI scanners, which are never turned off, often sit around unused at the weekends. So, with the help of friends and by circumventing the hospital authorities, he finally got access to such a machine.

The special circumstances required that the test subjects fulfil three criteria: they had to be slim, flexible and not suffer from claustrophobia. Van Andel remembered his friends Ida and Jupp, who fitted the bill, and also, as street acrobats, were used to 'performing under stress', as the researchers put it.

Once they were in the tunnel of the MRI scanner, they were given instructions

The internal organs during sexual intercourse, as sketched by Leonardo da Vinci (c.1493).

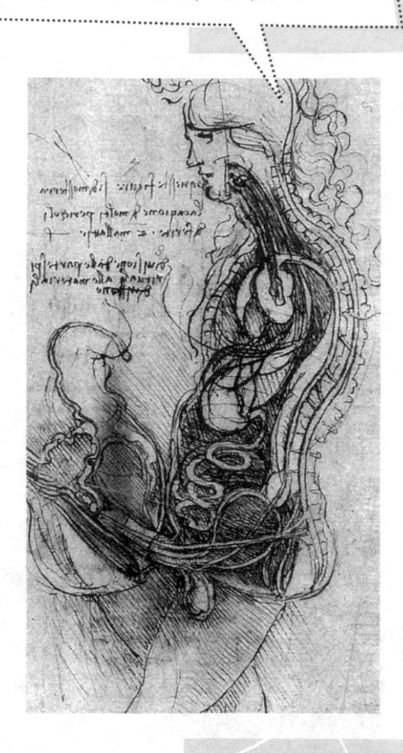

This longitudinal sectional drawing of the human anatomy during the sex act was published by Robert Latou Dickinson in 1933.

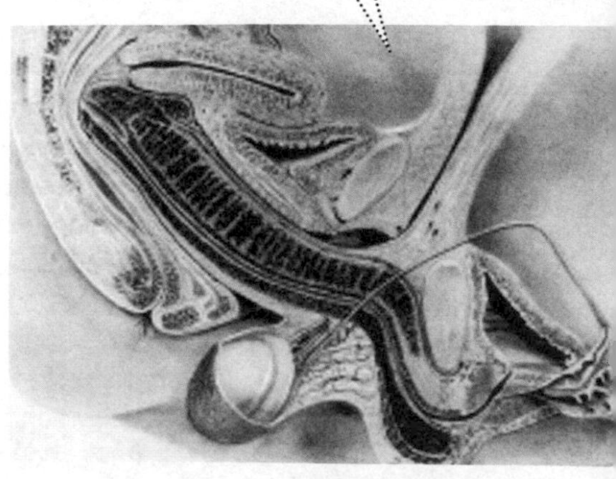

Longitudinal sectional image (magnetic resonance image) of the human anatomy during sexual intercourse (P = penis; Ur = urethra; U = womb; S = pubic symphysis; B = bladder; I = lower intestine; L5 = fifth lumbar vertebra; Sc = scrotum).

over an intercom system from the control room: 'The erection is fully visible, including the root,' a voice through the loudspeaker announced, and then added, 'Now lie down very still, holding your breath during the shot!'

A magnetic resonance imager can take sectional images of the human body. In contrast to an X-ray,

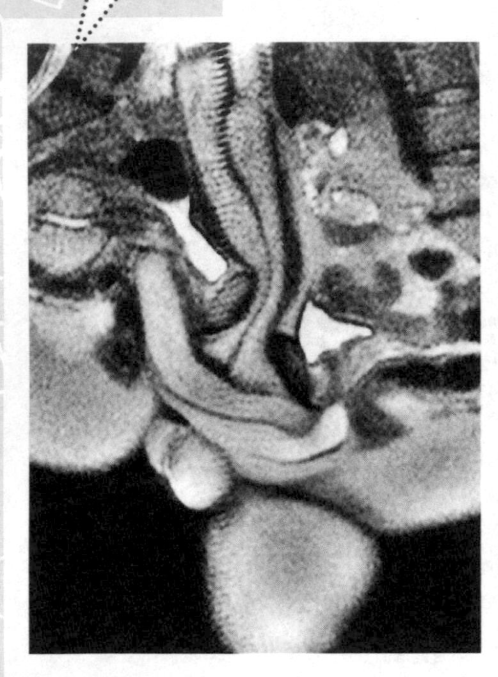

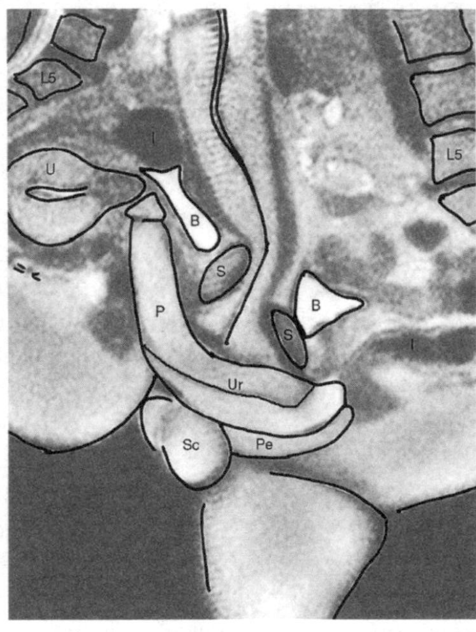

this technique doesn't put any strain on the body, but it does have one major drawback: the exposure time, during which the person being examined mustn't move, is long. In Jupp and Ida's case it was 52 seconds, while later experiments in a better machine took 12 seconds.

The paper detailing the experiment was rejected three times by publishers. The editors of some medical journals weren't sure whether someone was playing a joke with a spurious article. The *British Medical Journal*, on the other hand, asked for more data, stating that it was impossible to draw scientific conclusions from just a single couple.

Accordingly, van Andel solicited more test subjects. His appeal on a local television station caused quite a stir, sparking heated debates. It did, though, finally yield eight pairs and three single

madsciencebook.com

The Ig-Nobel Prize is awarded for examples of bizarre research. The list of past laureates can be seen at http://www.improbable.com/ig/ig-pastwinners.html.

women who were prepared to take part.

In conjunction with the researchers, Ida Sabelis kept detailed minutes of the experiment, which give a good impression of the arduous nature of the task that the test subjects let themselves in for: 'After a preliminary image for positioning the true pelvis of the woman was taken, the first image was taken with her lying on her back (image 1). Then the male was asked to climb into the tube and begin face-to-face coitus in the superior position (image 2). After this shot – successful or not – the man was asked to leave the tube and the woman was asked to stimulate her clitoris manually and to inform the researchers by intercom when she had reached the preorgasmic stage. Then she stopped the autostimulation for a third image (image 3). After that image was taken the woman restarted the stimulation to achieve an orgasm. Twenty minutes after the orgasm, the fourth image was taken (image 4).'

Under these conditions, most couples didn't manage to meet the requirements of the experiment in the 20-inch tunnel: 'We did not foresee that the men would have more problems with sexual performance (maintaining their erection) than the women in the scanner.' In fact, Pek van Andel wouldn't have managed to get his pictures if Viagra hadn't come on the market in Holland in 1998. After several failed experiments, two of the men took this drug to increase their potency. An hour later, things clicked.

The image that the *BMJ* finally published in its Christmas issue of 1999 shows that the root of the penis comprises one-third of its total length and, contrary to expectations, takes on the form of a boomerang in the missionary position.

The experiment earned the authors an 'Ig-Nobel prize', and provoked a heated debate among the specialists. One doctor even suggested resorting to porn stars for future experiments: 'These people are especially trained to complete the sexual act under all types of conditions.' Another questioned how meaningful the data were, since in the narrow tunnel, the women couldn't spread their legs sufficiently to achieve the 'true missionary position'.

Source

Schultz, W.W., et al. (1999), 'Magnetic Resonance Imaging of Male and Female Genitals During Coitus and Female Sexual Arousal'. *British Medical Journal* 319, pp. 1596–1600.

"The erection is fully visible, including the root"

1994 Fair-weather Waiter

In March 1994 there was a room-waiter in one of the casino hotels in Atlantic City, New Jersey, who provided a strange weather forecasting service. Every morning, before knocking on a guest's door with breakfast, he would take a deck of cards out of his tunic and pick one of them. On it was written one of four forecasts: 'cold and rainy', 'cold and sunny', 'warm and rainy', or 'warm and sunny'. The hotel rooms were soundproofed and fitted out with darkened glass, making it impossible to see what the weather was like from inside. If the guests asked what it was like, the waiter, irrespective of the actual weather conditions, would always tell them what he'd read on the card and then leave the room.

This experiment was devised by the American psychologist Bruce Rind to find out whether not only real weather conditions but also the anticipation of particular weather could affect a person's mood.

The tips that the waiter got indicated that 'direct sensory exposure to sky conditions is not required in order for these conditions to affect overt behaviour'. In other words, it was in the waiter's interest to lie if the weather was poor. In this case, he got about a third more in tips if he told the guests that the sun was shining. The temperature was irrelevant.

For waiters who are uncomfortable with lying to someone's face, here's the good news: four years later the same researcher discovered in an Italian restaurant that even a good forecast for the following day jotted on the back of the bill yielded 25 per cent more in tips.

Source

Rind, B. (1996), 'Effects of Beliefs About Weather Conditions on Tipping'. *Journal of Applied Social Psychology* 26 (2), pp. 137–147.

1995 The Minimum Distance for Striptease Artistes

Is the state responsible for infringing the right to freedom of expression of exotic dancers if it forbids them by law from stripping entirely naked during their performances in nightclubs? This bizarre question was resolved between 19 and 23 August 1995 by an experiment that took place in the 'Little Darlings' club in Las Vegas.

In the USA freedom of expression is guaranteed by the First Amendment and is inviolable. Any ruling that attempts to infringe upon a freely expressed opinion, whatever that might be, is unconstitutional. However, the ban on naked dancing and regulations concerning the minimum distance between dancers and spectators were deemed by a number of courts not to be unconstitutional because they did not substantially alter the erotic message being conveyed by the dancer.

Daniel Linz and some of his colleagues at the University of California found this assumption to be somewhat unrealistic, and so travelled to Las Vegas to prove the opposite. One week before the experiment, eight dancers from the 'Little Darlings' were trained by a choreographer to strip off their black dresses exactly 30 seconds after the start of their performance. In the experiment that followed, they wore either bra and panties underneath or nothing, as the case may be. After the three-minute show, 24 test subjects (men between the ages of 18 an 65) filled out a questionnaire on the effect it had had on them. Statistical evaluation of the results 'revealed a significant difference between the nude and non-nude conditions on Erotic Communication ... the patrons in the nude condition are more likely to be receiving an erotic message than patrons in the non-nude condition'.

Source

Linz, D., et al. (2000), 'Testing Legal Assumptions Regarding the Effects of Dancer Nudity and Proximity to Patron on Erotic Expression'. *Law and Human Behaviour* 24 (5), pp. 507–533.

1997 Pubic Hairs Go Walkabout

The authors of the work 'Frequency of Pubic Hair Transfer During Sexual Intercourse' were at pains to leave their readers in no doubt about the motives of their test subjects. The paper by the forensic scientist David L. Exline and his colleagues stated clearly that, 'the only participation incentive was altruistic advancement of research'.

Six couples – employees of the Alabama Department of Forensic Sciences and their partners – were asked to take hair samples ten times after sexual intercourse, and, what's more, to adhere strictly to the 'standard pubic hair combing protocol' while doing so. This involved the test subjects laying out a 3-ft (90-cm) square paper towel under their partner's bottom and carefully combing through their pubic hairs so that any loose hairs fell onto the towel. They then had to place the towel and the comb inside an envelope and attach a questionnaire to it detailing the duration of intercourse, the length of time since their last bath and their last sexual contact, and the sexual positions they adopted.

In the resulting 110 hair samples – one of the couples supplied only 5 each – the scientists found 344 pubic hairs, 20 body hairs, 7 hairs from the head and one animal hair. At least one of the partner's hairs was found in 19 of the samples, which represented a transfer frequency of 17.3 per cent, with hairs migrating noticeably more often from the woman to the man (23.6 per cent) than the other way round (10.9 per cent). There was only one case of cross-transference, where hairs moved simultaneously from the man to the woman, and vice versa.

According to the researchers, this low transfer rate meant that sex offenders would be unlikely to be identified on the basis of their pubic hairs.

Source

Exline, D.L., et al. (1998), 'Frequency of Pubic Hair Transfer During Sexual Intercourse'. *Journal of Forensic Sciences* 43 (3), pp. 505–508.

1998 The Loudspeakers of Jericho

When the American educational television network The Learning Channel set about trying to get to the bottom of certain ancient mysteries from the Bible, top of its list were the trumpets of Jericho. The Book of Joshua describes how seven priests blew their trumpets in front of the Ark of the Covenant and made the walls come tumbling down. The Swiss UFO-logist Erich von Däniken doubted the lung capacity of the priests, and suggested that the walls might in fact have been brought down by some kind of sophisticated acoustic noise-generating device.

This spurred the television producers to commission Wyle Laboratories in California to build a small brick wall in their labs and assault it with sound waves from the largest loudspeaker they could find.

This task was tailor-made for Wyle's special WAS 3000 speaker, which emits a sound as loud as 10,000 normal home hi-fi speakers. After six minutes of sustained noise from this, the mortar did indeed begin to crumble and the wall collapsed. The producer of the show, Jim McQuillan, pronounced this result 'conclusive'. However, he didn't mean that von Däniken was right,

Source

Wyle Laboratories (1998), 'Wyle Completes Unique Tests in Investigation of Biblical Mysteries for Television Program' (press release).

but simply that the experiment confirmed the common wisdom '... that sound can, in fact, cause destruction'.

McQuillan was wise not to stick his neck out any further, for it has long been known that the Canaanite towns, which included Jericho, weren't fortified, meaning that there weren't any walls for the seven priests and their 10,000 hi-fi speakers to bring tumbling down in the first place.

> "that sound can, in fact, cause destruction"

1999 Unaccountably Hungry

Hunger is a constant source of mystery to scientists. Take, for example, the experiment set up by Barbara J. Rolls of Pennsylvania State University.

In her lab, Rolls served up similar starters to three groups of women. The first group were given a ragout of chicken, rice and vegetables, while the second got exactly the same dish, but this time turned into a soup, simply through the addition of 356 grams of water. Although this didn't change the energy content of the meal one iota – water has no calorific value – the soup was far more filling: those who had it as a starter ate a good quarter less of the main dish.

The greater volume of the soup went part of the way to explaining this result, but the truly bizarre outcome was what happened with the third group. They were given the ragout and a glass containing precisely the same volume of water that the second group had had blended into their soup. Therefore, within the same allotted time (12 minutes) these two groups ate an

Served up in different ways, the same basic ingredients with the same calorific count had a markedly different effect on how the person eating them felt.

identical amount and type of food, yet the soup eaters were far less hungry afterwards. Once again, they consumed 25 per cent less of the main course.

Source

Rolls, B.J., et al. (1999), 'Water Incorporated into a Food but not Served with a Food Decreases Energy Intake in Lean Women'. *American Journal of Clinical Nutrition* 70, pp. 448–455.

Even Rolls, who is known worldwide as one of the leading researchers into appetite, was at a loss to explain this behaviour. Her best guess was that just the sight of the soup had a more filling effect, since it took up more space in the bowl than the ragout.

Rolls's experiment demonstrated how little scientists really know about regulating hunger – and showed soup to be the glutton's enemy.

2002 The Mathematics of Stick-throwing

One October day in 2002, near the town of Holland on Lake Michigan, a man could be seen playing a curious game with his dog. Standing on the shoreline, he repeatedly threw a tennis ball diagonally into the water. The dog immediately went after the ball, while the man set off in pursuit of the dog. The dog raced down the beach for a stretch before jumping into the water. At that point, the man quickly stuck a screwdriver in the sand, grabbed the end of a tape measure that he'd laid out ready beforehand a bit further on and then also plunged into the lake after the ball. Over three hours, this extraordinary spectacle was repeated more than 40 times.

Using a tape measure and a screwdriver, Elvis's master discovered that the dog could solve a complex mathematical problem instinctively.

The man was called Tim Pennings, and he is a maths professor at Hope College in Holland, Michigan. The strange business with the screwdriver and tape measure was designed to answer the question of whether his dog Elvis could count. What's more, he wasn't trying to find out if Elvis could do simple multiplication, but instead solve a far more complex mathematical problem.

Source

Pennings, T.J. (2003), 'Do Dogs Know Calculus?'. *College Mathematics Journal* 34 (3), pp. 178–182.

To cut to the chase: Elvis definitely can count. Indeed, the whole world bore witness to the fact after Pennings was interviewed by the BBC, invited onto a talk show in Hollywood and even quoted in a Vietnamese newspaper.

Pennings had the idea for the experiment shortly after getting Elvis in August 2001. On their walks he'd keep throwing a tennis ball into the water for the dog, which dutifully fetched it. Pennings noted; 'As I was watching Elvis race down the beach and then plunge into the water just before arriving at the perpendicular, it suddenly occurred to me that he was taking a path exactly as I sketch out in my calculus classes when I show my students an optimization problem.' In the actual problem, Tarzan has to run along a riverbank before swimming across it to rescue Jane from quicksand on the opposite side. The question was exactly where he should jump into the water in order to reach Jane as quickly as possible.

Like Tarzan, Elvis is faced with a number of possibilities. He could jump into the water immediately and make straight for the ball. That's certainly the shortest route, but not the fastest, since Elvis can't swim as fast as he can run. Elvis could also run along the beach until he is immediately opposite the ball before jumping in. This would reduce the swimming leg to a minimum, but at the same time make the entire distance as long as it possibly could be. The quickest route lies somewhere between: first run along the shore for a bit and then swim diagonally to the ball. The ideal point at which to start swimming is dependent upon the relationship between swimming and running speed.

By throwing a tennis ball into the water for him, Pennings wanted to find out whether Elvis really could solve this optimization problem correctly. To do so, he first determined Elvis's speed on land and in the water: his running speed was 21 ft (6.4 m) per second, and his swimming speed 3 ft (0.91 m) per second. From this, Pennings could pinpoint the optimal place at which Elvis should go from running to swimming. And, lo and behold, in almost every test, Elvis hit the spot.

Was Elvis really able to solve this complex mathematical problem? Pennings was the first to quell any hype on this score: 'We confess that although he made good choices, Elvis does not know calculus. In fact, he has trouble differentiating even simple polynomials.' Rather, it was clearly the case that he hit on the best solution quite intuitively.

At the end of his paper on the subject, Pennings suggests repeating the experiment with a dog that has to decide when to leave a snow-free path and plunge into a deep snowdrift. Or doing the same thing with six-year-olds, primary-school kids or students. However, 'for the sake of their pride, it might be best not to include professors in the study'.

2003 Close Encounters With a Dogbot

In the early days of behavioural research, scientists would try to fool animals with crude dummies (see p. 154). In the interim, however, the robot age has also caught up with ethology. Researchers from the Eötvös Loránd University in Budapest and the Sony Computer Science Laboratory in Paris tried to find out whether dogs would accept Sony's commercial dogbot AIBO as one of their own. And so they brought 40 dogs into contact with the techno-dog, which measures 1 ft (30 cm) from nose to tail and weighs in at just over 3 pounds (1.5 kg). For some of the tests, they covered AIBO in a fleece that they'd put in a puppy's basket the day before.

Cutting-edge behavioural research: the dogbot AIBO encounters a flesh-and-blood dog.

As a control, the scientists also showed the dogs a real puppy and a toy car. From their orientation to and distance away from the dogbot, and from the frequency with which they barked, growled and sniffed the dogbot (its front and rear ends), they concluded that 'at present there are some serious limitations in using AIBO robots for behavioural tests with dogs'. The dogs did react to the dogbot, but far less strongly than they did to the puppy.

madsciencebook.com Watch a film showing just how badly a dogbot can fare when it meets a real dog.

On their website, the researchers are at pains to point out that no animals were harmed in the course of the experiment. AIBO was on the receiving end of several attacks, but kept on working like clockwork. Even so, the researchers warn you not to try this at home: the manufacturer's warranty on the AIBO doesn't cover this kind of damage.

Source

Kubinyi, E., et al. (2004), 'Social Behaviour of Dogs Encountering AIBO, an Animal-Like Robot in a Neutral and in a Feeding Situation'. *Behavioural Processes* 65, pp. 231–239.

Acknowledgements

This isn't just my book. But for the help and generosity of a whole host of people, these pages would have remained blank. First and foremost, I'd like to thank the scientists who conducted the experiments, and who agreed either to meet me or to give me the lowdown on their work over the 'phone or by e-mail. Many of them delved into their archives and dug out unpublished papers for me, or conjured up photos that had long been thought lost.

In scouting around for source material, I relied on the kind assistance of many people. Bernd Wechner sent me several studies on hitch-hiking that had come to him via Richard Pomazal. Even the authors of these studies didn't have copies of them any more. Peter Cogman gave me some invaluable tips on how to go about finding French studies on the guillotine. Sasha Andreev-Andrievsky helped me with my research in Moscow, while my thanks are due to Wladimir Bitter for his translations from Russian. Meanwhile, in the USA, Andreas Rüesch, Christine Andres und Gaudenz Danuser gave me access to their correspondence.

After her work on this book, my research assistant Stella Martino can fairly claim to be a specialist in offbeat scientific journals, as can Kathrin Hoffmann in the realm of acquiring bizarre scientific images. Urban Fetz, a wizard on the scanner, managed to make photographs from old magazines sharper than they ever had been in the original. Thomas Häusler, Urs Willmann, André Schneider and Daniel Weber read the manuscript, in whole or in part, and saved me from committing many factual errors and stylistic clangers.

My agent Peter Fritz got behind this book right from the word go and gave it his wholehearted backing.

My colleagues in the editorial department at *NZZ Folio* – Lilli Binzegger, Andreas Dietrich, Andreas Heller and Daniel Weber – gave me the benefit of their wealth of experience in journalism, Ernst Jaeger was quick on the draw on the scanner, while Esther Baumann sustained me with her delicious chocolate cakes. I'd always imagined my dream job to be just like this.

Peter Lewis did a wonderful job translating the book into English, and Slav Todorov at Quercus was extremely forbearing in dealing with a welter of e-mails from me with last-gasp minor changes.

As well as reading through the entire manuscript, my wife Regula von Felten also had to put up with repeated mealtime discussions about resuscitated dogs' heads or sweat-gland extract in doctors' waiting rooms. Long may our experiment together continue!

Index

Figures in italics indicate illustrations.

Picture Acknowledgements

P. 10, 11: Basle, Public Library of the University of Basle, Mscr. F IV 30
P. 13: Image courtesy of the Blocker History of Medicine Collections, Moody Medical Library. The University of Texas Medical Branch at Galveston, Galveston, Texas
P. 17: Wellcome Library, London
P. 18: from: L'ettres sur L'electricité, dans lequelles on trouvera les principaux phénomènes qui ont été découverts depuis 1760, avec des discussions sur les conséquences qu'on en peut tirer, Paris, 1767
P. 20: Wellcome Library, London
P. 23: Bibliothèque Nationale, Paris
P. 25: ETH-Bibliothek, Zürich (from: Aldini, J. Essai théorique et expérimental sur le Galvanisme: avec une série d'experiences faites en présence des commissaires de L'Institut National de France, et en divers amphitéatres anatomiques de Londres, Vol. 1&2)
P. 28: Library of Congress, Washington DC
P. 29: Wellcome Library, London
P. 33: Museum Boerhaave, Leiden, Netherlands
P. 36, 37: École Nationale Supérieure des Beaux-Arts, Paris
P. 40: Reproduced by kind permission of Alan G. Ingham
P. 44: Wellcome Library, London
P. 45: from: Aminoff, M.J. (1993) *Brown-Séquard: a visionary of science*, Raven Press, P. 168
P. 47: Cinémathèque Française collections des Appareils, Paris
P. 51: Reproduced by kind permission of the University of California, Berkeley, Department of Psychology
P. 54: Wellcome Library, London
P. 56 top: from: Munn, Norman L., *Handbook of Psychological Research on the Rat: An Introduction to Animal Psychology*, 1950. Houghton Mifflin and Company, Boston:
P. 56 bottom: CartoonStock/Jim Sizemore
P. 57 top: *The New Yorker* Collection 1994 Sam Gross of Cartoonbank.com. All rights reserved
P. 57 bottom: The New Yorker Collection 2002 Mike Twohy of Cartoonbank.com. All rights reserved
P. 61 top: from the *Washington Post*, Mar. 18, 1907
P. 61 bottom: from the *Washington Post*, Mar. 12, 1907
P. 62: *Focus Features*, Los Angeles und New York
P. 64, 65 top: Wellcome Library, London
P. 65 bottom: Cartoonstock/Parolini + Elmer
P. 66: Nightmare Records
P. 67, 68: from: Karl, K. (1912) *Denkende Tiere: Beiträge zur Tierseelenkunde auf Grund eigener Versuche*, Engelmann
P. 69 top: Bildarchiv Preußischer Kulturbesitz, Berlin
P. 69 bottom: from Kladderadatsch, 1909
P. 70: from the Stevens Point Daily Journal Oct. 13, 1904, Stevens Point, Wisconsin
P. 73 top, 73 bottom: Lederle Labs
P. 74 top: from the *Reno Evening Gazette*, Jan. 17, 1922
P. 74 bottom: Wellcome Library, London
P. 76 : (c) Berlin-Brandenburgische Akademie der Wissenschaften (Berlin—Brandenburg Academy of Sciences; formerly the Prussian Academy of Sciences)
P. 78: Prof. Ben Harris, Department of Psychology, University of New Hampshire
P. 83: from: Finkler, W. Kopftransplantation an Insekten. Archiv für mikroskopische Anatomie und Entwicklungsmechanik, 1923, pp. 104–33
P. 86: Reproduced by kind permission of the AT&T Corp., USA
P. 87: Reproduced by kind permission of the Harvard University Archives
P. 92: Title image from 'Science and Invention', May 1927
P. 93: from: Dickinson, R.L., *Human Sex Anatomy*. 2nd edition 1949 (1st edition 1933), Williams & Wilkins, Baltimore
P. 96, 97: from: Kraus, J.H., 'The Living Head', in *Science and Invention*, 1929, pp.922–3
P. 98: B.F. Skinner Foundation
P. 99 top: (c) *The New Yorker* Collection 1993 Tom Cheney of Cartoonbank. com. All rights reserved
P. 99 bottom: (c) www.Perspicuity.com
P. 102: Stanford University News Service
PP. 103–7: from: Kellog, W.N. and L.A., *The ape and the child*, 1933, Hafner Publishing Company, New York
P. 108: Getty Images/Time Life Pictures
P. 110: Corbis/Bettmann Collection
P. 111, 112: Getty Images/Kirkland
P. 115: Getty Images/Time Life Pictures
P. 116: from the Iowa City Press Citizen, Nov. 14, 1946
P. 121: Das Fotoarchiv/SVT Bild
P. 122: Getty Images/Hulton Archive
P. 124: Getty Images/Hulton Archive
P. 125: NASA
P. 133: NASA
P. 135: from: *Human Behavior*, Time Life Books 1976,
P. 137: *Lethbridge Herald*, Dec. 16, 1954
P. 138 top: ITAR-TASS/P. Khorenko and Yu. Mosenzhnik
P. 138 bottom: Reto U. Schneider, Zürich
P. 141: Shurley, J. T., Profound experimental sensory isolation. Amer. J. Psychiat., 1960, Vol 117: p. 539-545. Reproduced by kind permission of the American Psychiatric Association
P. 142: Warner Home Video
P. 143: Corbis/Ressmeyer
P. 144: J. Tapprich, Zürich
P. 146: Reproduced by kind permission of U.S. Army, Fort Detrick, Maryland
P. 150: Reproduced by kind permission of Fort Lee Film Commission, New Jersey, USA
P. 151: from the *New York Times*, May 19, 1958
P. 154, 155: University of Wisconsin Archives, USA
P. 156: from the *Stevens Point Daily Journal*, Nov. 15, 1958
P. 157: from: Ward, J.E. Physiologic Response to Subgravity I. Mechanics of Nourishment and Deglutition of Solids and Liquids. *Journal of Aviation Medicine*, 1959, 30: p.153
P. 159: Associated Press
P. 160: Reproduced by kind permission of Harvard University Archives
P. 163: from the Herold-Press, Jan. 29, 1960
P. 168: Collection of Alexandra Milgram. Reproduced by kind permission of Alexandra Milgram
P. 169, 170: Stills from the film '*Obedience*', 1965 by Stanley Milgram and distributed by Penn State Media Sales. Reproduced by kind permission of Alexandra Milgram
P. 171: Pandemonium Records
P. 176: Getty Images/Hulton Archive
P. 181: from: *Psychology Today*, Vol. 3, No. 3 (June 1969)
P. 183, 184: from: Delgado, J.M.R. *Physical control of the mind: toward a psychocivilized society*. 1969, Harper & Row, New York
P. 188: (c) 1998 by James McMullan. Reproduced by kind permission of Penguin/Penguin Group USA Inc.
P. 193: Stanford University News Service
P. 196: Reproduced by kind permission of P.G. Zimbardo Inc.
P. 203: Reproduced by kind permission of Karl Heider
P. 207: Reproduced by kind permission of J.M. Darley and C. D. Batson
P. 213: Reproduced by kind permission of John Ware
P. 215–219: Reproduced by kind permission of P.G. Zimbardo Inc.
P. 220 top: World Domination Music Group, Hollywood, California
P. 220 bottom: P.G. Zimbardo
P. 225, 226: Associated Press
P. 230, 231, 233: Reproduced by kind permission of Scherz Verlag, Zürich/Santiago Genovés
P. 232: from the *News Journal*, May 5, 1973
P. 234: Dan Heller, Vancouver, Canada
P. 236: NASA
P. 240: from: Baron, R. A. The reduction of human aggression: a field study of the influence of incompatible reactions. *Journal of Applied Social Psychology*, 1976, 6: pp. 260–74
P. 243: Reproduced by kind permission of Jürgen Klapprott
P. 244: from the *New York Post*, Mar. 3, 1978
P. 246: Wolfgang Krüger Verlag
P. 249: NASA
P. 258: Reproduced by kind permission of Ben Libet
P. 265: Reproduced by kind permission of Barry Marshall
P. 267: Bildagentur Focus/SPL, Hamburg
P. 269: Reto U. Schneider, Zürich
P. 271 top: Leonardo da Vinci, Longitudinal section of a man and woman during coitus, c.1492
P. 271 bottom: from: Dickinson, R.L., *Human Sex Anatomy*, 2nd edition 1949 (1st edition, 1933), Williams & Wilkins, Baltimore
P. 272: Prof. W. W. Schultz/*British Medical Journal*
P. 277: Reproduced by kind permission of Barbara Rolls
P. 278: Reproduced by kind permission of Tim Pennings
P. 280: Reproduced by kind permission of Kubinyi Enikõ

First published in Great Britain in 2008 by
Quercus
21 Bloomsbury Square
London
WC1A 2NS

Originally published in Germany in 2004 by Bertelsmann
under the title *Das Buch der Verrückten Experimente.*

A CIP catalogue record for this book is available from the British
Library.

ISBN 978 1 84724 494 9
10 9 8 7 6 5 4 3 2 1

Printed in Great Britain by Clays Ltd, St Ives plc